T0296169

THE PROGRESS OF PHYSICS

Yours truly,
J. Clerk Maxwell

THE PROGRESS OF PHYSICS

DURING 33 YEARS (1875—1908)

FOUR LECTURES DELIVERED TO THE UNIVERSITY
OF CALCUTTA DURING MARCH 1908

by

ARTHUR SCHUSTER, F.R.S.

Ph.D. (Heidelberg), Sc.D. (Cantab.), D.Sc. (Manchester and Calcutta),
D.ès.Sc. (Geneva)

Cambridge:
at the University Press
1911

CAMBRIDGE
UNIVERSITY PRESS

University Printing House, Cambridge CB2 8BS, United Kingdom

Cambridge University Press is part of the University of Cambridge.

It furthers the University's mission by disseminating knowledge in the pursuit of education, learning and research at the highest international levels of excellence.

www.cambridge.org
Information on this title: www.cambridge.org/9781107559905

First published 1911
First paperback edition 2015

A catalogue record for this publication is available from the British Library

ISBN 978-1-107-55990-5 Paperback

THIS VOLUME IS DEDICATED

TO

THE HONORABLE Mr JUSTICE
ASUTOSH MUKHOPADHYAYA, SARASWATI,

C.S.I., M.A., D.L., D.Sc.

VICE-CHANCELLOR OF THE UNIVERSITY OF CALCUTTA,

IN ACKNOWLEDGMENT OF THE SERVICES
HE HAS RENDERED TO UNIVERSITY
EDUCATION IN INDIA

INTRODUCTION

IN writing out these lectures which were delivered nearly three years ago, I have made a few additions, in order to bring the subjects to their position at the present time. Personal opinions have been influenced in some cases by subsequent researches or by maturer reflexion. All such expressions of opinion must therefore be considered as indicating my present attitude rather than that taken up at the time of the lectures. Their scope and aim has been sufficiently indicated in the opening passages.

ARTHUR SCHUSTER.

December 1910.

TABLE OF CONTENTS

x   *Table of Contents*

LECTURE IV

The author is indebted to the proprietors of *Nature* for permission
to insert, as frontispiece, a copy of the portrait of J. Clerk Maxwell,
which appeared in that journal in the series of "Science Worthies."

LECTURE I

It is more than thirty years ago, that returning from a scientific expedition I hurriedly passed through India, and fascinated by its sunshine, the beauties of its mountain scenery and the mysteries of its national life, carried home with me the intense desire of a more leisurely visit. I then stood at the beginning of my scientific career and when now, nearing its end, my wish is fulfilled by being called upon to fill the honourable office of " Reader " to the University of Calcutta, I thought it might be interesting to you—as it is to me—if I discharge my duties by reviewing the progress of scientific thought in the interval.

It is my aim to trace the changes in our point of view, rather than to give an historical account of the sequence of the discoveries which make this period memorable. In taking this course I am aware of the risk incurred in an attempt, which to a great extent must be guided by subjective impressions ; for though a man may see clearly the changes that have taken place in his own mind, he may be wrong in estimating the rise of the average level of scientific thought by the standard of his individual progress. According as he has been taught by advanced or antiquated teachers, according as he ends his work ahead of, or behind his time, will he give a very different version of the progress which has been achieved. But rightly or wrongly I have chosen my course : I prefer to be frankly subjective, and warn you beforehand that my account will be fragmentary, and to a great extent reminiscent of those aspects which have come under my own personal view.

Let us briefly survey the general position of physical science at the time our story begins. The work of the previous twenty years was dominated by the gradual recognition of the Science of Energy founded experimentally on the work of Joule, and theoretically on that of Kelvin, Clausius and Helmholtz. Destined from the beginning to serve as a bond between all branches of physical knowledge, it found its first triumph in establishing the connexion between the science of heat and the theorems of dynamics. The first law of thermodynamics, which teaches us that heat is generated or destroyed in exact proportion to the amount of mechanical work lost or gained, was not only recognised as unassailable, but had truly saturated the soul of the scientific body. Such was not entirely the case (is it now ?) with the second law, which regulates the conditions, under which heat can be changed into mechanical work. It is true that every one whose opinion counted, accepted the law, and no student was allowed to remain ignorant of Carnot's cycle, and of the mathematical equations which can be derived from it ; but the law was not a living and fertilising part of physical science ; except in its special applications to heat engines. In its more refined consequences dealing with entropy and available energy it had not yet become common property, and in its application to the decay of universal energy it was not sufficiently understood and appreciated.

The tendency to hide ignorance under the cover of a mathematical formula had already appeared but was not openly advocated ; hence students were still taught to form definite ideas of the processes of nature. In the kinetic theory of gases ample material was found for a mental picture of the state of matter in its simplest form. The calculation of the average molecular velocities, the application of the theory to the processes of diffusion, conduction and viscosity had been accomplished and was taught at some, at any rate, of the Universities. I became acquainted with the theory of gases in lectures given by L. Soret at Geneva in 1869, and well remember how the possibility of determining the size of molecules and their number in a given volume seemed to open out before me, but though I attacked the problem I could make nothing of it, and it was not till some time later that I found it already solved by Johnstone Stoney and Loschmidt. Those who could follow the more intricate mathematical developments of the subject found in Maxwell's investigation on the law of distribution of molecular velocities a new revelation, which introduced for the first time the theory of probability into problems relating to the physical state of bodies. In the succession of new ideas that have influenced the progress of Physics this great step must always hold a pre-eminent position.

As regards light, the elastic solid theory held the

field, though serious difficulties, passed over too lightly in the period of its triumph, already foreshadowed its ultimate defeat.

Stokes by his brilliant investigations had placed the laws of transmission of vibrations through elastic solids on a firm basis. He had shewn how a complete theory must take account of two waves, the condensational wave and the distortional one. The first of these is important in the theory of sound, but no optical phenomenon indicates its existence and hence arises a certain difficulty. When a single homogeneous medium is considered, this difficulty is most easily overcome by assuming the condensational wave to be transmitted with infinite velocity, but Lord Kelvin shewed that consistent results are obtained equally well, by introducing the condition that these waves are propagated with a velocity which is infinitely small. This however was at a date considerably later than the one to which I am at present referring. Condensational waves were then generally ignored, but another stumbling block worried our minds. The laws of transmission of light through different bodies could, up to a certain point, be equally well explained by assuming the elasticity of the æther to be the same everywhere, and its density to be different in different bodies, or by specifying that the density is constant, while the elasticity changes. An increased velocity inside a body could accordingly either be due to an increased elasticity or to a diminished density. The relationship between the direction of vibration in a beam of light and what is called its plane of polarisation depends on which of the two alternatives we adopt. If the density be variable, the direction of vibration must be at right angles to the plane of polar-isation, while parallelism between the direction of vibration and the plane of polarisation would signify that the elasticity differs. The more refined conse-quences of the theories gave rise to the hope, that experiments might be devised to decide between the two alternatives, but unfortunately the verdict of ex-periment was not uniformly on the same side. When Fresnel's formula for the intensity of a ray reflected from a transparent body, which had been verified experimentally by Brewster, was closely criticised by Lord Rayleigh, its theoretical foundations were only found to be sound, on the supposition of variable density. This meant that the vibration was at right angles to the plane of polarisation, a result which was also supported by the polarisation observed in a wave scattered by small particles.

But the law of double refraction in crystalline media led to the opposite conclusion. Here again Fresnel was the pioneer, who gave us the equation of the wave surface by specifying that elasticity in a crystal depends on the direction of displacement. Physicists were now placed in the dilemma, that in different parts of optics two mutually destructive theories on the properties of the æther were supported by experiment. For though Rayleigh shewed that we may obtain a surface not differing very much from Fresnel's, by assuming a variable inertia according to the direction of vibration, the difference was sufficient to admit of the question being decided by experiment, which, performed by Glazebrook gave its verdict emphatically in favour of Fresnel. These difficulties and discrepancies, though not considered fatal, prepared men of science for the great revolution, which was soon to come and sweep away the whole elastic solid theory.

The state of electrical science in 1870 must present itself somewhat differently to the scrutiny of a historian who derives his information from the records of published papers, and to the recollection of a student, who received his impressions from the teachers at the time. Maxwell's great paper "A Dynamical Theory of the Electromagnetic Field" appeared in 1864, but I doubt whether the younger generation of physicists had their attention drawn to, or seriously arrested by it, before the publication in 1872 of the two volumes *Electricity and Magnetism*. I believe that the first systematic course of lectures based on Maxwell's theory was given by myself at the Owens College during the session 1875–76. (Sir Joseph Thomson was one of the three students attending the course.) At the time Maxwell's volume appeared, the teaching of electricity centred round the calculation of coefficients of induction and futile discussions on the laws of action of so-called current elements. A wider and more philosophic aspect was now brought before us, and though Maxwell's treatise was essentially mathematical, it put an end to the university tradition of looking upon electricity as part of applied mathematics to the neglect of its physical aspect. Hence Maxwell's work changed the whole point of view of the study of electrical science.

While the above indicates in brief outline the general condition of the main branches of Physics at the period of which we are speaking, we must now turn our attention to the important question as to how students were prepared by teaching and example to take an active part in advancing their subject.

I think I interpret correctly the recollection of those who passed through their scientific education at the time, when I say that the general impression they received was that, apart from theoretical work, a reputation could only be secured by improved methods of measurement which would extend the numerical accuracy of the determination of physical constants. In many cases the student was led to believe that the main facts of nature were all known, that the chances of any great discovery being made by experiment were vanishingly small, and that therefore the experimentalists work consisted in deciding between rival theories, or in finding some small residual effect, which might add a more or less important detail to the theory. There were no doubt great differences of opinion depending on the temperament of the teacher, as to how far increased accuracy of measurement was an object desirable in itself or only a means to an end, and in this connexion a passage taken from Clerk Maxwell's Introductory Lecture in Experimental Physics deserves to be quoted[1]:

" This characteristic of modern experiments—that they consist principally of measurements—is so prominent, that the opinion seems to have got abroad, that in a few years all the great physical constants will have been approximately estimated, and that the only occupation which will then be left to men of science

---

[1] *Collected Works*, II. p. 241. No date is attached to this lecture in the published volume, but according to the information given in Maxwell's *Life* it was delivered in October, 1871.

will be to carry on these measurements to another place of decimals.

"If this is really the state of things to which we are approaching, our Laboratory may perhaps become celebrated as a place of conscientious labour and consummate skill, but it will be out of place in the University, and ought rather to be classed with the other great workshops of our country, where equal ability is directed to more useful ends.

"But we have no right to think thus of the unsearchable riches of creation, or of the untried fertility of those fresh minds into which these riches will continue to be poured. It may possibly be true that, in some of those fields of discovery which lie open to such rough observations as can be made without artificial methods, the great explorers of former times have appropriated most of what is valuable, and that the gleanings which remain are sought after, rather for their abstruseness, than for their intrinsic worth. But the history of science shews that even during the phase of her progress in which she devotes herself to improving the accuracy of the numerical measurement of quantities with which she has long been familiar, she is preparing the materials for the subjugation of the new regions, which would have remained unknown if she had been contented with the rough methods of her early pioneers. I might bring forward instances gathered from every branch of science, shewing how the labour of careful measurement has been rewarded by the discovery of new fields of research, and by the development of new scientific ideas. But the history of the science of terrestrial magnetism affords us a sufficient example of what may be done by experiments in concert, such as we hope some day to perform in our Laboratory."

Another interesting passage throws light on Maxwell's idea of the proper function of a laboratory.

" Our principal work, however, in the laboratory must be to acquaint ourselves with all kinds of scientific methods, to compare them, and to estimate their value. It will, I think, be a result worthy of our University, and more likely to be accomplished here than in any private laboratory, if, by the free and full discussion of the relative value of different scientific procedures, we succeed in forming a school of scientific criticism, and in assisting the development of the doctrine of method."

Maxwell's view was that of the most enlightened and progressive philosopher then living, but his optimism was not shared by others who at the time enjoyed an equal or greater reputation. Gustav Kirchhoff, for instance, was by no means a man who despised experimental enquiry; I have heard him speak with appreciation of men who without much theoretical knowledge tried to upset theory by experiment; but the sole advantage he expected to accrue from their labours was the mending of the theory and he did not anticipate new facts being discovered leading to a revision of fundamental conceptions. When I told him of the discovery then made in England, that light falling on the surface of a bar of selenium altered its electrical conductivity, he remarked: " I am surprised that so curious a phenomenon should have remained undiscovered till now." This represents the attitude of mind not only of Kirchhoff but of the great majority of physicists at the time. Looking back now on this period, when Roentgen rays and radio-activity were undreamt of, we may well learn to be cautious in our own predictions of the future.

It will be noticed that though Maxwell evidently looked forward to further discoveries, he expected them to appear as residual effects rather than by a direct experimental attack deliberately made to break new ground. A typical example of a great discovery brought about by the method indicated by Maxwell is furnished by Rayleigh's measurements, which culminated in the discovery of argon through a research undertaken to determine with accuracy the density of gases, and shewing unexpectedly a marked discrepancy between two samples of apparently pure nitrogen. The gas derived from air was found to be heavier than that generated chemically, because as was ultimately shewn, it contained small quantities of a denser and new gas. But though this may be quoted as an example of the orthodox method of discovery, and also fully recognising that a teacher is almost bound to point to this as the soundest method, it is nevertheless indisputable that the greatest discoveries, both formerly and in recent years have not originated in the hunt of residuals. Altogether I am doubtful whether any great discovery has ever been made by anyone who has only aimed at recording a number of facts. I do not believe, in spite of what is sometimes asserted, that Schwabe was led to the discovery of the periodicity of sunspots simply by sitting down to record without ulterior motive the number of spots he could see ; and it is to me unthinkable that the determination of the position of planets, which allowed Kepler to discover his laws, was made by men who, solely for their own pleasure or satisfaction, recorded their places among the fixed stars. We may be certain that Schwabe had some idea on the possibility of connecting sunspots with other phenomena, and we know that the old astronomers were not doing their work simply for the sake of curiosity, or in the vague hope that something unforeseen should come of it. Similarly we may be certain that Lord Rayleigh's object in determining the density of nitrogen, was not that of obtaining a more accurate value of a physical quantity without some definite view, why an increased accuracy in the determination of that constant was desirable. The most important result of a research no doubt is very often not that which was originally aimed at, but this does not prove the absence of an intention, which went beyond that of mere increased accuracy or an amplified record of facts.

My object is not, however, to shew what students ought or ought not to have been taught, but what they actually were taught thirty years ago. In this respect Maxwell's declaration must be taken to represent the most progressive view of the time, but unfortunately only few students came under his direct influence.

I shall endeavour to describe the atmosphere which surrounded a student of Physics in the five institutions in which I worked. The method of testing a student's fitness for graduation differed then, as it does now, in different countries and naturally formed a potent factor in determining the course of instruction. In Germany the presentation of a thesis was almost universally a necessity, and consequently the Professor had to find some easy subject on which his pupils had to exercise their powers. There is no doubt that this system brings a student quickly to the limits of knowledge in one special department and he learns that science even in simple problems is progressive and unlimited. This is a decided advantage, counter-balanced however to some extent by the tendency it induces of exaggerating the value of small and unimportant matters. The problem of a thesis must be definite and limited. Unless a teacher is himself in the front rank of research, he cannot avoid suggesting some small piece of detailed work which may be useful as a training and even as a statistical registration of facts, but is apt to distort the student's perspective. The type of problem to which, as his first piece of research, he is apt to attach importance, has a tendency to affect his future aspect of nature, for the pleasures of studying minute detail are fascinating and grow quickly to a point where accuracy becomes an end in itself instead of a means to an end. In so far as the educational system is national do university requirements impress a national character on research work; but we must be careful not to mistake this for an inborn racial difference in the manner of looking at nature.

At the time to which my remarks apply, a thesis in an experimental subject was often undertaken without any previous training in the laboratory such as is everywhere now introduced. Comparing the results of the old and modern system we are led to the conclusion that at present we probably attach too much importance to such preliminary training. Experience has shewn that some practice in the most common physical manipulation is desirable, but there is nothing to prove that a moderate time spent on a comparatively few typical experiments is not all that is required for the purpose. The need for some systematic training is acknowledged and had made itself felt everywhere and almost simultaneously in different countries. In Germany, the first edition of Kohlrausch's text-book appeared in 1869, and the practice of laboratory training in our present sense, spread slowly but steadily through all Universities. Kirchhoff, at Heidelberg, attached great importance to a carefully prepared scheme of observation accurately carried out. Once a week he gave a lecture explaining the principle of some experiment and the methods of calculation. As he generally had not more than twelve students, one morning or one afternoon in the week could be appropriated to each of them, to perform the manipulation which had been described in the previous lecture. The student, at the time set apart for him, found the apparatus ready but only partially adjusted, and without further assistance was expected to complete the adjustment and obtain an accurate result. Once during the time of work Kirchhoff came round to see whether the student had got into serious difficulty. The answers had to be written out and were carefully entered up in a book. At the beginning of a subsequent lecture, they were all written up on a blackboard, and the professor commented on their agreement or disagreement, and discussed generally the accuracy which could be

expected in any particular case. In addition to
this so-called " Seminar," which only occupied part
of the year, a few men were admitted to the laboratory
to carry on research work. In the two semesters
over which my stay extended, Lippmann was completing
his celebrated research on the capillary electrometer, and
Kammerlingh Onnes was preparing his doctor's dissertation, on a modified form of Foucault's pendulum.
I had told Kirchhoff that my main object was to
learn from him as much as I could, and that I did
not mind whether any definite result came out of my
work or not. He consequently set me to test some
instrument he had designed to determine the elliptic
polarisation of light reflected from a metallic mirror ;
a typical piece of work of the orthodox stamp. The
instrument was not a particularly successful one, and
there was nothing to publish when it had been tested,
but I learned something both theoretically and experimentally on elliptic polarisation and I was quite
content. In addition to a very elementary course on
Experimental Physics, Kirchhoff gave some theoretical
lectures which were of the greatest interest. He
was a very precise man who weighed every word he
said, and was commonly reported never to have
missed a lecture[1]. Whenever a student ventured
on any remark respecting some question in Physics

[1] I can speak however from personal knowledge of an exception.
" Gentlemen," he told us at the end of a lecture, "I regret to inform
you that circumstances compel me to omit my lecture to-morrow."
The circumstances referred to were, that he was going to get married.
The omitted lecture was on a Saturday, Sunday was reserved for the
wedding tour and on Monday the lectures proceeded as usual.

he had to be prepared to stand a cross examination on the precise meaning of what he said—but with all that, Kirchhoff was one of the kindest of men, who judged others leniently, though he expected the same standard of accuracy in others which he had set for himself.

During the summer of 1874 I spent a few months at Göttingen, the recollection of which is valuable to me chiefly on account of the scientific intercourse I had with Wilhelm Weber. That name probably conveys but little meaning to the students of the present day, yet two generations ago he was a leader of science, and one of the founders of our present system of electric units and measurement. Weber's laboratory, housed in a private residence, carried the distinction that the first electric message transmitted along wires was sent from it to the astronomical observatory, then in charge of Gauss.

The contemplation of that period takes us back far beyond the time which I am now trying to recall, and I will not dwell upon it. Weber, when I knew him, was 70 years old, and had preserved an active and elastic mind. I was endeavouring to find some effects of the material through which an electric current was passing on the intensity of that current in violation of Ohm's law. Looking up the literature, I found that this law rested on very slender experimental basis, and my experiments actually seemed to indicate a deviation, observable when an alternate current was superposed on a direct one. Weber entered with great spirit into the question, though the result apparently shewed a breakdown of the law, in the opposite direction from that which he anticipated in consequence of his own theoretical speculations. My experiments ultimately led to an investigation by a British Association Committee, resulting in a vindication of Ohm's law to a very high degree of accuracy[1]. The directorship of the laboratory of Göttingen had, shortly before my arrival, been handed over by Weber to Riecke, who still holds it. There was only one student at work beside myself; for the purposes of his doctor dissertation he was magnetising ellipsoids and testing magnetic formulæ in the orthodox fashion.

I now turn to describe a laboratory of very different character and ambitions. Helmholtz, who then stood at the height of his power, naturally drew the most promising students from all parts of Germany to Berlin, where they worked in the two or three rooms, which then constituted his laboratory. Most of them were preparing their doctor dissertation on some subject directly rising out of Helmholtz' work. As examples of such work, in which the generally limited and sometimes moderate powers of a student were utilised to the fullest extent to advance science, I may mention Moser's work on the electromotive forces set up between two solutions of the same salt in different states of concentration, and E. Root's verification of the penetration of hydrogen into platinum under the influence of electric polarisation. Needless to say, no efforts were made in this laboratory to push numerical measurement beyond its legitimate limits, and though most of the work done was quantitative in character, qualitative

---

[1] The experiments were designed by Maxwell and performed by Chrystal.

experiments were not discouraged. Goldstein was working at his electric discharges in high vacua, and trying to explain effects he discovered near the kathode by a theory which I know to have been distasteful to Helmholtz ; yet no word did he ever say to discourage the purely experimental side of those experiments. I always considered Helmholtz' laboratory, such as it was at the time, the ideal of what a teaching laboratory should be. There were no doubt some inconveniences owing to want of space, each student having only a table at his disposal, but the simplicity and compactness of the arrangement allowed Helmholtz to make his round every day and to speak separately to each student. We profited not only by the advice given to us individually, but also by the suggestions made to our colleagues, and reaped the advantage of a detailed study of one problem without losing the wider culture of becoming interested in a number of different questions. The lesson I then learned in the conduct of a laboratory was still further impressed upon me during a visit which I paid to Berlin a few years later, when a new laboratory had been built under Helmholtz' own direction. In place of a small number of badly furnished and crowded rooms, there was now a noble building, impressive on the outside and perhaps also inside to the casual visitor, but all the soul and scientific spirit of the old place had gone. The laboratory was constructed on the principle so dear to beginners, that every one should have a private room in which he could set up his apparatus without fear of outside inter- ference. In consequence, Helmholtz no longer found the time personally to see and to advise the students, who worked without a common bond and generally without scientific impulse, and mainly for the purpose of completing a dissertation of sufficient merit for their degree. Experience to my mind shews that the system of closed compartments in a scientific laboratory, which is still adhered to in many cases, encourages secrecy and does not foster the true scientific spirit in the community. It is significant that Helmholtz did not long retain the directorship of the "Institut," but resigned shortly after its opening.

The two laboratories at Berlin such as I knew them represent two extremes, but how is the designer of a new building to avoid the inconveniences and combine the advantages of both? How is he to give increased space, better equipment, greater facilities of research, without loss of that contact between the students, and between the student and professor which is the essence of the educational value of a laboratory? The problem is difficult, but not impossible to solve, if we bear the main objects of practical instruction in mind.

Though its educational value cannot be called into question, it is doubtful whether the systematic training in practical work, which has grown up during the last twenty-five years, has had a material influence on the progress of science. In entering into the question of laboratory organisation, I am therefore laying myself open to the charge of wandering from my subject; but in appreciating the different agencies that have been at work to increase knowledge, we cannot altogether ignore one, on which strong hopes were built, and which, if it has hitherto failed, partially or completely, has done so because it has not yet found its right methods. We may hope to remedy this most quickly by facing the facts : and the cardinal fact to bear in mind is that previous to 1870, when laboratory instruction, if given at all, was sporadic, the experimental skill of investigators was as great as it is now. We need only mention the names of Faraday, Joule, Helmholtz, and Regnault, in support of that contention. The conclusion is irresistible that an intelligent student possessing a sufficient theoretical knowledge, is capable of starting research work in Physics without previous special training. It is not for him that complicated laboratory courses have been designed, but for the ordinary student, who requires a certain amount of theoretical knowledge, which is acquired far more easily, if the principal phenomena are brought to his direct notice by practical work. The future investigator will no doubt ultimately save time, if at an early stage he acquires a certain technical skill, and becomes acquainted with physical methods ; but otherwise the efforts of the teacher should be directed to stimulating his scientific activity rather than sending him through all manipulations which he might possibly have to perform. If we place this stimulation of mind in the foreground, we must all the more insist on the necessity of securing an intimate intellectual intercourse between students and Professors. Accepting this as the correct view, nothing can be more antagonistic to the true interests of a student than his condemnation to solitary confinement in a cell, while he is performing his first research. This tendency of laboratory designers to favour solitary work may be partly responsible for the small number of experimental discoveries which can be traced to the direct influence of laboratory teaching.

The organisation of laboratory work in England has to some extent been controlled by examination requirements, though in the early part of the period which 'I am considering, there was no such thing as laboratory examination. Nevertheless, the more rigid method of instruction which resulted from the control of examination in other departments, unconsciously affected practical teaching, and led more quickly to a systematic organisation with all its advantages and drawbacks. Where the number of students to be taught is small, all organisation can be dispensed with, for the student is most quickly taught to depend on his own resources when the teacher is still tentatively feeling his way either in consequence of his own immatureness or the unsettled state of his subject. At any rate, I felt in my own case the advantage I derived from beginning my practical training at a time when there were no text-books on the subject. Balfour Stewart introduced laboratory teaching in the old buildings of Owens College soon after his appointment in 1870. A terrible railway accident which nearly cost him his life interrupted his work, and only in the autumn of 1871, when I entered the college, was he able to resume it. Though Stewart was always ready to help us in all serious difficulties, the main part of the instruction was handed over to teachers, whose knowledge was tempered by sufficient ignorance to allow them to be dogmatic without carrying conviction. In consequence we—the students—not being able to distinguish between bluff and solid teaching, took nothing on trust and started investigations of our own, to confirm or disprove our instructors. To cite one case, I remember being extremely puzzled by finding that in measuring the refractive index of a glass prism, two widely separated images of the sodium line appeared. I thought of internal reflexions as an explanation, but the position of the lines did not support this and I ultimately consulted one of the teachers, who in a superior tone rebuked me for not knowing that the sodium line was double. He had not noticed that there was an angle of several degrees between the components and one of the images, though yellow was standing right over the blue part of the continuous spectrum accompanying the other component. In conjunction with a fellow student I pursued the enquiry, and ultimately we found in some French book that unannealed glass was doubly refractive. It seems that the prism we experimented with, was originally purchased to shew this double refraction, and to the end of my tenure of the professorship I shewed it annually to illustrate the effect.

Balfour Stewart, whose work on the Theory of Exchanges holds a high position in the history of Physics, was an inspiring teacher because he was one of the few who did not discourage attempts to discover new facts. He himself was always busy trying to open out new fields of enquiry. In a room partly used for laboratory instruction, he experimented with a disc set into rapid rotation in vacuo, with a view to finding a thermal effect. He was led to try this, because he could not reconcile in his mind the continuance of thermal equilibrium with the change in the radiation and absorption, which in consequence of the Doppler effect would result when the velocity of a body changes. Balfour Stewart experimented from the point of view of one, who thought that a breakdown of the second law of thermodynamics was conceivable when new conditions were introduced. Wien at a later period was led by the same idea, but basing himself on the truth of the second law, he deduced the relationship between the intensities of radiation for different periods which is consistent with that law, when the bodies within the enclosure are in motion. He thus made an important addition to the science of Radiation.

I came into closer contact with Balfour Stewart when I returned to Manchester in 1873 as Demonstrator of Physics. His mind was then running on the possible discovery of a minute variation of gravity due either to chemical combination or to a screening effect which might shew itself when the weight of a disc is determined with its plane first in the vertical and then in the horizontal position. He also made a number of experiments to discover some kind of interference between two crossing electric currents, as when currents are made to pass at right angles to each other from two independent cells through an electrolyte. In none of these experiments was he successful, but the students benefited from his alertness and freshness of mind. My own ideas at the time were dominated partly by Weber's explanation of diamagnetism, through electric currents circulating round molecules and partly by the first impressions I received on reading Maxwell's *Theory of Electricity*. I formed some idea that if light is an electrodynamic disturbance, Weber's currents might be brought into connexion with it and consequently tried a number of experiments (described in a Note Book which is still in my possession) on the influence of magnetism on the vibrations emitted by a sodium flame. All this was only vaguely before me, but I placed a sodium flame in the strongest magnetic field which the laboratory could provide and looked at the result through a spectroscope. To my delight I found a widening of the lines. I sometimes wonder what would have been the result if I had then published what might now be considered as an anticipation of Zeeman's discovery; but as a matter of fact the effect was spurious and with the spectroscopic resolving powers used, I could not possibly have seen the real effect. After a few days of exciting work, I found indeed that the steel spring of the slit was drawn aside whenever the magnetic field was excited, and that the broadening of the line was simply due to a widening of the slit. By arranging the experiment differently, I could produce a narrowing of the line in place of the broadening. An experience such as this ought to shew the danger of taking it too easily for granted, that an observation recorded in some old book is an anticipation of a later discovery, when in reality it is probably only the result of a careless experiment accidently simulating an effect which is real, but can only be detected by more refined methods.

The opening of the Cavendish Laboratory at Cambridge marks the beginning of a new era of physical discovery; for though the results achieved in actual research work were small during the first years of its existence, the indirect effect of the laboratory

and of Maxwell's personality on Cambridge thought
was very considerable. Nor was this influence confined to Cambridge, for it was Maxwell's writings
which inspired Hertz in the work which laid the
foundation of a new school of physics in Germany.

The Cavendish Laboratory owes its existence to
the generosity of the Duke of Devonshire, who was
Chancellor of the University, and as Chairman of a
Royal Commission on Scientific Education had been
impressed by the need of institutions in which experimental research could be carried on. The building
was formally opened on June 16th, 1874. Lord
Kelvin (then Sir Wm Thomson) had declined the
new Chair of Experimental Physics, and after some
hesitation, and only in consequence of considerable
pressure put upon him by his scientific friends, Clerk
Maxwell consented to be nominated. The formal
appointment was made on March 8th, 1871. The
introductory lecture, of which I have already quoted
a significant passage, shews us the ideal which was in
Maxwell's mind, but Universities do not alter their
habits very rapidly and the whole Cambridge system
of education did not fit in well with the new development. The qualification for graduation on the scientific
side was dominated by the Mathematical Tripos, which
covered a range of studies too wide already to admit
of time being available for additional instruction in
experimental work.

Physics looked at from the theoretical point of view
always held a position and occasionally an important
one in the Tripos, though the policy of the University
with regard to that subject was not always consistent.

In 1849, the mathematical theories of electricity, magnetism, and heat, were definitely excluded from the examination, and capillary attraction soon followed, but all these subjects were re-introduced in 1873. The detachment of the theoretical study of Physics, as it then was conducted at Cambridge, from the experimental presentation of facts, led in some cases to a remarkable antagonism to ocular demonstration which is illustrated by a characteristic incident. Clerk Maxwell who possessed an innate desire to see what he could with his own eyes, had taken considerable trouble in cutting and grinding a plate out of a doubly refracting crystal to shew conical refraction. The experiment is difficult, and delighted at its successful accomplishment Maxwell met one of the mathematical teachers of the University. "Would you like to see Conical Refraction?" asked Maxwell. "No," replied Todhunter, "I have been teaching it all my life, and I do not want to have all my ideas upset by seeing it." That this remark was not made jocularly is shewn by a passage which occurs in an essay on the *Conflict of Studies*, in which Todhunter discusses the advisability of introducing experimental illustrations into the lessons given in schools. He declares himself as opposed to it on the ground, that an experiment which is not intended to bring out a new fact is useless, and proceeds as follows: "It may be said that the fact makes a stronger impression on the boy through the medium of his sight, that he believes it the more confidently. I say that this ought not to be the case. If he does not believe the statements of his tutor—probably a clergyman of mature knowledge, recognised ability, and blameless character—his suspicion is irrational, and manifests a want of power of appreciating evidence, a want fatal to his success in that branch of science which he is supposed to be cultivating."

Todhunter was a highly cultivated man and an able mathematician; he undeniably believed in the mission of a University to advance knowledge. "There appear to be," he writes, "three distinct functions of the University. One is that of examination, one that of teaching, one that of fostering original research: the first of these three has practically been as yet most regarded, and many of us hope that it will in future decline either absolutely or relatively by the increased development of the other two[1]." Or again, consider this passage :—

"Of that public professional instruction which is often regarded as the essence of a University, there is comparatively little in Cambridge. During the last quarter of a century out of the whole range of mixed mathematics lectures have been regularly delivered by professors only on the following subjects: Optics, Hydrostatics, Astronomical Instruments and Lunar Theory. I have myself sometimes received letters of enquiry from strangers who wished to study certain branches of pure or mixed mathematics, and contemplated spending a year or more at the University for the advantage of professional lectures. It was not quite satisfactory to be compelled to reply that there was scarcely any of that machinery for teaching of which the applicants seemed to assume the existence."

[1] *Conflict of Studies*, p. 238.

While Todhunter believed in advancing knowledge by experiment, he attached no value to one which is made to improve the mind of the student. His frame of mind was that of one satisfied with reading about foreign countries in books, believing that the traveller who goes to see them, only incurs trouble and expense while he would gain morally by staying at home and shewing his faith in the original explorer. It would be wrong to take Todhunter as the type of a numerous class ; he was a freak who differed from his type in having the courage of his opinions. It is not surprising, considering the manifold conflicting interests which must always retard the progress of a University, that the Cavendish Laboratory only slowly obtained its grip on Cambridge education ; that it obtained it as quickly as it did was a sign of the rapidly growing influence of men animated by the true progressive spirit which spread through the University, and is the cause of the pre-eminent position which Cambridge now holds in the world of science.

During the spring of the year 1876 I determined to give up the position I held at Owens College, in order to work a year or two at Cambridge : the time of my stay there was gradually extended until my final return to Manchester in 1881. I was well received by Maxwell, but otherwise was looked upon with friendly curiosity by a University that was not accustomed to see anyone coming to it merely for the purposes of research, without previously availing themselves of the benefits of degree courses and the prospect of the usual University rewards. After a little trouble and delay I was allowed to enter one of the colleges.

My position was irregular, if not illegal, for though inscribed as an undergraduate I was allowed to stay five years without passing the "Little Go," though subject, of course, to the discipline of the Proctor. I remember meeting that formidable officer of the University at dinner, and walking home with him unchallenged and not fined, though I was without cap and gown. Research students were only recognised at a much later date, and though their numbers are now large, I feel sorry that it should not have been possible to attract them by the advantages to be gained through working under a leader of science rather than through the bribe of a degree.

The Cavendish Laboratory presented a different aspect then, from that which it assumed at a later time when students began to fill its rooms. As far as I remember, William Garnett, who acted as demonstrator, and George Chrystal, now Professor of Mathematics at Edinburgh, were the only regular workers. W. M. Hicks, now Professor of Physics at the University of Sheffield, came occasionally; but shortly afterwards R. T. Glazebrook and W. N. Shaw joined us, and we began to consider ourselves crowded. The facilities of the laboratory were not what students now expect. We had to charge our own batteries and learn a little glass blowing and ordinary workshop manipulations, as there were no instrument makers nearer than London. But every one who worked in the Cavendish Laboratory during these last years of Maxwell's life must retain for ever the impression of Maxwell's intellectual influence and charm of manner. Except when prevented by his own ill-health or that of his wife, he visited the laboratory daily and made the round of the rooms in which any work was going on. He might ask a question about the progress of one's experiments, but more generally spoke about the subject which occupied his thoughts at the time, for he was much absorbed in his own ideas and not always quick in bringing his mind to bear on a fresh subject. It would occasionally happen that he took no notice of a question put to him, leaving it doubtful whether he had heard it or not, but next day he would probably start his conversation with: " By the way, there was a question you asked me yesterday and I have been thinking about it...." Then would follow a suggestive and fully considered answer. All who were brought into contact with him must remember some of the quaint and humorous sayings which gave such a charm to his conversation. His thoughts at the time seemed to run much on what is now called the equi-partition of energy. Boltzmann's researches had recently been published, and Maxwell seemed to accept them, though only with hesitation, because he could not see how far they might lead us. I remember especially his saying that Boltzmann's theorem if true, ought to be applicable to liquids and solids as well as gases.

Maxwell took an intense interest in the publication of the papers of Cavendish which he was then editing, and with great enthusiasm repeated most of the experiments described in them. He seemed specially fascinated by the manner in which Cavendish anticipated subsequent discoveries made with delicate instruments by converting himself into a galvanometer. No effects of electric currents known to Cavendish could serve for the measurement of their strength, and he therefore estimated the intensity of the physiological shock which he felt when a current was suddenly sent through his body, the experiments being arranged so that two currents were declared to be equal, when the shocks they produced appeared to be equally intense. Every one who came to the laboratory at that time had to submit himself to a sometimes unpleasant test, in order to convince himself that the method could give consistent results.

It is impossible to say whether Maxwell was satisfied with the position the laboratory took in the first years of its existence. It grew steadily in importance, but the progress was slow and required time to develop, but time—alas—was not vouchsafed to him. Maxwell died in the autumn of 1879 at the age of 48. His ideals soon began to be realised under the professoriate of his successor, Lord Rayleigh.

Systematic instruction which now became necessary in the preparation for the Natural Sciences Tripos was excellently supplied by R. T. Glazebrook and W. N. Shaw, who acted as demonstrators. For the purpose of encouraging research, Lord Rayleigh planned to take up a problem requiring extensive series of measurements, in which a number of students could take part, who would thus gain experience and at the same time the power of combination which is only possible in an organised laboratory would be utilised. His first choice of subject, the redetermination of the electrical unit of resistance was a most appropriate one, because the revolving coil of the British Association which had served for the first determination of the Ohm was available in the laboratory, and the accuracy of this determination had been called in question by Rowlands work. To its shame, be it said, no one in the University was willing to help, and it was only to assist a good cause and to set a good example that I regretfully left my own experiments to take part in the redetermination of the Ohm. I have never regretted, however, being associated with Lord Rayleigh in the early stages of the work in which Mrs Henry Sidgwick soon joined, and which proved to be of permanent value to electrical science.

If I have entered, perhaps too much, into a discussion of laboratory organisation, it was because many hopes were founded on the introduction of practical work as an essential portion of degree courses. There is no doubt that the general standard of knowledge has been raised through the more intimate contact of a student with the facts of nature ; it is also open to us to believe, though it may be difficult to prove, that the progress of knowledge as tested by the discovery of new facts was accelerated.

It is time, however, that we should return to our main object, which is to trace the outline of knowledge before the great revival of experimental discovery. I have already indicated the position of the main branches of Physics, but there were a few detached portions in which experiment was ahead of theory, and which to some extent were destined to form a connecting link between the old and the new aspect of science. The discovery of spectrum analysis and its application to the study of the chemical composition of celestial bodies, kept alive our faith that man had not reached the end of his endeavours to bring new facts to light, and we began to recognise that the radiations of a luminous gas were likely to give us valuable information on molecular structure. The first simple conception that the spectrum of an element was characteristic of an atom, and accompanied that atom through its various combinations with other bodies, was rudely disturbed when it was found that compounds had their own spectra, and that even an elementary body could under different circumstances possess more than one spectrum. This last fact was not generally admitted without opposition, extending through several years, because it is always difficult to answer objections based on a supposed impurity of the material used. The facts however soon began to be too strong for the doubters, and multiple spectra established their position. The longer spectroscopy was studied, the greater its complications appeared, especially in the case of gases rendered luminous under reduced pressure. To clear up the difficulties it became necessary to examine more closely the discharge itself, and this cleared the way for one of the great advances of which I shall have to speak in a subsequent lecture.

An example of an experimental discovery arising out of irregularities observed in the course of accurate numerical measurements is to be found in Crookes' experiments, which led to the construction of the radiometer. Wishing to determine accurately the atomic weight of thallium, Crookes designed a balance placed in a receiver from which the air could be removed by means of an air pump. When the pressure had been reduced sufficiently to eliminate all effects of air currents, certain anomalies were observed indicating an apparent repulsive force acting on bodies exposed to heat radiations. An experiment made by myself proved that the forces depended on the presence of the residual gas still left in the receiver, and Crookes himself was ultimately able to improve the vacuum sufficiently to diminish and finally to destroy the effect. These experiments, which roused considerable interest at the time, take their place in the history of science not so much because they have given us an instrument which has been useful in the study of radiation, but because they shewed the imperfections of the so-called chemical vacua, which were then believed to contain no traces of air. As soon as this was recognised to be wrong, methods of removing the air from a vessel, rapidly became more effective.

Our brief account of the state of science when the Cavendish Laboratory was founded, would not be complete without some reference to a fundamental theory of matter, which absorbed a good deal of the energy of Cambridge mathematicians, and though now abandoned and neglected, will always stand as a monument of a great effort. When the hydrodynamical investigations of Helmholtz had shewn the possibility of constructing a universe in which the ultimate units were hydrodynamic entities, Lord Kelvin with characteristic energy developed a theory, in which the atom was considered to be a vortex ring in an incompressible fluid, and this theory fascinated and satisfied the mind because it went to the very

foundation of the edifice of nature.   Kirchhoff, a man
of cold temperament, could be roused to enthusiasm
when speaking about it.   It is a beautiful theory, he
once told me, because it excludes everything else.
For the present the theory is set aside, but may yet
be revived in a different shape.   In one respect its
point of view resembles that of later theories, because
it recognises only one substance, and makes the energy
of a moving atom and therefore its mass, depend on
the kinetic energy of the surrounding medium.   The
study of electrical phenomena has led to the same idea
by a different road, and it will be my duty in the
next lectures to shew how a more perfect representation
of the ultimate constitution of ponderable matter is
foreshadowed by the more recent electrical discoveries.

LECTURE II

WHEN we consider two rival theories in any branch of human knowledge, we are sometimes drawn towards one or towards the other, not by any process of reasoning, but by an instinctive feeling which may be so strong, that we unhesitatingly reject one of the alternatives. We flatter ourselves in such cases that our choice is guided by convictions gained in previous studies, and is therefore based on a sound though unconscious application of the reasoning faculty. This use of an instinct, which we like to call "common sense," is justified when applied to one portion of a subject, that has already been examined from a wider point of view; but can we trust to a mere bias—which may be a prejudice as well as a true instinctive guide—in deciding a question of fundamental character?

If I open my second lecture with this psychological query, it is because I should like you to consider briefly what justification there is for that strong instinctive feeling, which denies action at a distance. By action at a distance we understand, that two bodies can act on each other, without being connected, by some medium which transmits the force.

The denial of action at a distance implies that we only recognise forces, which directly push or pull the body that is to be moved.   There can be no question that, as an article of faith, the "push and pull" theory of force has dominated the scientific thought of the last half century ; and though there are signs of a coming reaction there is still an overwhelming preponderance of feeling in favour of an almost axiomatic negation of action at a distance.   It would be interesting to trace in detail the causes of such firm belief in the two mutually exclusive doctrines that "a body cannot act where it is not," and that "two bodies cannot occupy the same space," but we must content ourselves with a few brief reflections.   To me it seems that the dogmatical denial of action at a distance is a survival of the ancient anthropomorphic explanation of natural phenomena.   Bodies are moved on the surface of the earth by the pulling or pushing of living bodies, and we therefore unconsciously associate a sense of muscular effort with the force that causes motion. When a body drops down from a height, without any apparent cause, we instinctively assign to its fall a cause essentially identical with that which can lift it up again ; for to connect two similar effects with similar or identical causes, is the most elementary instinctive principle to which probably all others may be reduced. Is not the fundamental instinct, which makes action at a distance seem unnatural to us, identical with that which placed the sun in a chariot drawn by horses in order to explain his motion across the heavens ?   The formulation of the principle of contact forces as an act of faith is modern, but our pride in it may to some extent be chastened, if we recognise its want of originality.

Let us look at the question a little more closely. When we push a body with our hands, we take the effect to be an example of a force acting by contact and not "at a distance." But if our senses were much more acute than they are, and we could distinguish by sight the separate molecules of a body, we should probably see no contact anywhere, the molecules being always separated by small intervals, and our conclusion in witnessing the effect of the pressure exerted by our hand would be that the body that is pushed begins to move away before the molecules of our hand have come into contact with it. Our primitive feeling then would be in favour of action at a distance, which shews that our present instinctive conviction against it, may only be due to defects of our senses.

If I am anxious that you should realise how weak are the grounds on which we deny on principle action at a distance, it is only to lay the greater stress on the real advance which science has made in consequence of the belief that all action between two bodies is transmitted through a connecting medium. Two lessons may here be learned. One is, that the temporary success of a doctrine does not necessarily justify the grounds of its foundation, and the other that progress in science is more often achieved by a definite hypothesis, which may be followed up and tested, than by a wider and perhaps more philosophic doctrine, which cannot be disproved, because it does not endeavour to go deeper than the mere descriptive classification of phenomena. At present, we may take it, that the doctrine of " no action at a distance " has secured its successes, mainly because it opened out a new field, which could be brought to the test of experiments. These successes must be placed to the credit of that school of science which explains the unseen by means of models such as we can construct with material bodies.

It is interesting to trace the gradual development of scientific ideas on the properties of the medium which transmits the vibrations of light. We must, for this purpose, go back to the foundation of modern astronomical science by Sir Isaac Newton. The simplicity and success of his treatment of gravitation tempted philosophers to a similar proceeding in investigating the problems of molecular forces. The ultimate constituents of matter came therefore to be looked upon merely as centres of force and the elastic properties of solids had to be deduced from the properties of such centres. When it was realised that the vibrations of the luminiferous æther are similar to the vibrations of an elastic medium, the æther was endowed with an atomic constitution, and its power to transmit transverse undulations had to be traced back to central atomic forces. This was a logical proceeding, but it failed because it took scientific men too far beyond what could be ascertained by observation and experiment. The next step restricted the problem, in accordance with the general experience, that the difficulty in going ahead does not lie so much in making a step sufficiently large to constitute a real advance, as in finding one small enough to be taken with safety. If the æther behaves like an elastic body, it is not necessary to make the investigation more difficult by enquiring what it is that gives a body the property of elasticity, but it is sufficient to deduce the phenomena of light, postulating as an assumption that the æther possesses the elastic properties of known solid bodies. The elastic behaviour of a finite portion of a body may be studied without reference to any molecular forces, and assuming these properties to persist when that portion becomes indefinitely small, we may calculate the effects of disturbing forces which are applied to any portion of the body. The result will be substantially correct, so long as the assumed elastic properties hold for portions of the body as small as the length of the wave.

It was the great work of Green and Stokes to have treated the problem of light as one of elasticity, by assuming no properties of the æther beyond those which we know elastic solids to possess. The merit of this work is by no means diminished by the ultimate failure of the theory, for its failure could only be established by the strict investigation of its consequences.

While the elastic solid theory of light was being developed by the great mathematicians of seventy years ago, Faraday's genius gave us a new and lofty standpoint for the outlook on electrical phenomena. Whether his treatment of electric action, as transmitted through a medium, permeated the scientific mind and affected contemporary thought, is difficult to decide, but at any rate it convinced Maxwell, and that was sufficient.

The strongest of our scientific "instincts" is our ultimate belief in the simplicity of nature. If both

light and electrical attractions are transmitted through
a medium, it would revolt our feelings—that is to say
our non-reasoning faculties—to assume, without strong
evidence to the contrary, that two different media exist
for the two manifestations.   But while it was possible
to explain light by means of a medium having the
properties of bodies known to our senses, this medium
failed at the same time to explain electric actions.
Maxwell's genius inverted the problem. He asked
himself what the properties of a medium should be in
order to account for the transmission of electrodynamic
effects, and he then discovered that such a medium
would transmit waves with a velocity of propagation,
exactly equal to that with which light was known to
be transmitted.   Hence any medium which could explain electrical action could also explain light.   This
is Maxwell's greatest achievement and the foundation
of the electrodynamic theory of light.

Although I may go over ground that is familiar to
many of you, I shall briefly explain the main features
of Maxwell's theory, because I am anxious to point
out, that success was achieved mainly through the
clearness and definiteness of the physical conception,
on which the mathematical development was based.

Steady electric currents, if the current be defined
in the usual way, satisfy the laws of flow of an incompressible fluid.   If, for instance, water flows down a
river bed which may be wide in some parts, narrow
in others, steep or nearly level in different places, the
volume of water which crosses each section in a certain
time is the same everywhere.   Hence, where the river
is deep or wide the velocity of the water must be corre-

spondingly less than where it is shallow or narrow. The same law holds in the case of electric currents which are permanently maintained : the same amount of electricity flows through every part of the circuit. This suggests as a first hypothesis that a current of electricity means the flow of something which *always* behaves like an incompressible fluid. The essential generalisation here lies in the word "always," which extends what we observe in the case of certain electric currents, where the observation can be carried out, to all classes of currents, some of which differ very materially from those, for which the proof can be given. Such generalisations are characteristic of the usual procedure of scientific discovery when we apply what can be proved in special cases to the general case and see whether the wider proposition leads to contradictions. To give, as an example, the case of a flow of electricity, which seems to violate the conditions of an incompressible fluid, imagine an electric battery E (Fig. 1) and connect its poles N and P to the plates of a condenser, A and B. While the condenser plates are charging, an electric current passes through the battery and the connecting wires. According to the older view, while electricity thus accumulates on A and B, there is no current through the intervening space. But if the flow of electricity is always like that of an incompressible fluid, the current cannot stop at the condenser plates, but must be completed through the space between them. There should there-

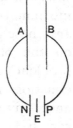

Fig. 1.

fore be an electric current through the non-conducting space between the plates *A* and *B* when the battery is connected, but only while the plates are charging. When this is done there is no further current through any part of the circuit. The question which now arises, is whether the current which can temporarily pass through a non-conducting medium has all the properties of an ordinary electric current, more especially its electro-magnetic properties. Maxwell assumes that it has, and this carries with it the consequence, that a change of magnetic force causes an induced current in a non-conducting as well as in a conducting medium. Here then we have a generalisation which is sufficiently precise to allow of its being followed up mathematically and being confirmed or disproved by experiment.

If at some point of space a change of electric force takes place, as for instance when an electrified conductor be moved backward and forward, there must, if Maxwell is right, be electric currents in that part of the medium which immediately surrounds the conductor; and these again must act by electro-magnetic induction on the other adjacent parts of the medium. The problem is quite definite and leads to a system of equations for the propagation of an electrical disturbance, giving a constant rate of propagation, which can be calculated. The numerical data which are necessary for the purpose are obtained by comparing, e.g. in the above example (Fig. 1) the electro-magnetic force of the charging current with the electrostatic attraction between the plates, when fully charged. It is found that the velocity of propagation of electro-magnetic effects is equal to that of light. Hence we can calculate the velocity of light, making use only of purely electric measurements. This is convincing, and leaves no reasonable ground for doubting, that light and electrical actions are both transmitted through the same medium : that light is in fact an electro-magnetic phenomenon.

In his first endeavour to develop Faraday's theory of electric action, Maxwell went beyond the necessities of the case, so far as the electro-magnetic theory of light is concerned. The scientific period of which I am speaking was essentially one of model making. Physicists still felt the necessity of having clear con-ceptions, and distrusted mathematical symbols that had no fully defined meaning. If the æther be called upon to transmit electric and magnetic actions, scientific men felt uncertain of their ground, unless they had some notion how the necessary stresses and strains were brought about. Hence Maxwell, who unquestionably belonged to this school, which requires a mechanical model for the conception of physical phenomena, devoted a good deal of time to the specification of electrostatic and magnetic strains in his medium. It is not necessary to enter into that part of his work here, because the question has for the moment been pushed into the background ; nor is it necessary to enter into the successes, which the electro-magnetic theory soon achieved by overcoming the difficulties of the older theories of light.

The issue on which Maxwell's results went consider-ably ahead of experimental knowledge was the trans-

mission of electro-magnetic effects with finite velocity.
Everybody knows that the position of a delicately suspended magnetic needle is affected by a magnet which
is placed in its neighbourhood, and if this magnet be
shifted the needle will in general turn to one side. So
far as our senses can judge, the action is instantaneous
and the needle turns at the moment the magnet is
displaced. But Maxwell says: No! not at the same
moment but after a certain interval of time only,
though this interval is too short to be apprehended by
our senses. If the distance of the magnet be one
metre the time of transmission of the effect would be
the three hundred millionth part of a second. Can we
obtain an experimental verification of this time of
transmission?

Surprise may reasonably be expressed that while
Maxwell was surrounded at the Cavendish Laboratory
by a number of young physicists, who firmly believed
in his electro-magnetic theory, no attempt was made
by them to furnish an experimental proof of their
master's theoretical deductions. The explanation lies to
a great extent in Maxwell's habit of letting his students
go their own way and find their own problems. Unless
a student had asked him directly to suggest a problem,
I doubt whether it would have occurred to him to give
advice in the selection of a subject for investigation.
Nevertheless, and I speak from personal knowledge,
the desirability of an experimental proof of Maxwell's
theory was realised by Cambridge men, and other
British physicists, who were in contact with them, such
as FitzGerald. But the experimental difficulties seemed
formidable, notably as regards the emission of electro-

magnetic waves of sufficient intensity to give a measure-able effect at a distance. Had there been two equally probable theories in the field, I doubt not the attempt to carry out a crucial experiment would have been made ; but we were perhaps over confident in the inherent truth and simplicity of Maxwell's conception. An extended experimental investigation would cer-tainly have absorbed much time and labour which we did not consider worth undertaking, considering the indirect evidence in favour of the electro-magnetic theory, which seemed to make the result a foregone conclusion. We were wrong, because we forgot that the great body of scientific thought abroad, and to some extent in this country, was apathetic and even reluctant to abandon an elastically solid æther which had done good service, and to accept in its place a medium, the properties of which were unlike those of any known body. Even Lord Kelvin con-tinued to place the weight of his great authority on the scale of the older views.

In the meantime Helmholtz, always on the alert on the side of progress, suggested to his pupil, Heinrich Hertz, to take up the experimental investigation of the problem. Everybody knows the splendid manner in which Hertz accomplished his task : how he over-came the formidable experimental difficulties and succeeded in convincing the scientific world of the truth of Maxwell's theory.

The " Hertzian " wave carries a name which Hertz would have repudiated, but it commemorates a man who is associated with Maxwell in one of the greatest scientific achievements. This habit of attaching names to laws, physical methods, or instruments, where the law, method or instrument can be equally well described by some descriptive name, is not one that should in general be encouraged. Personal names are not merely stumbling-blocks to the student, but—and this is my main objection—they often establish false scientific history. In the present instance and without wishing to detract from the great merits of a personal friend I see no reason why "electro-magnetic wave" is not a sufficient designation.

Through the work of Hertz it was finally established that electro-magnetic effects, radiant heat and light are all transmitted through the same medium by disturbances which in all respects are identical, and only require different receiving appliances to render them apparent to our senses. The recording or observing instrument naturally has to adapt itself to the scale and type of the disturbance. Just as we should use different instruments, according as we wish to record the tidal change of level of the ocean, or the waves engendered by a storm or the ripples on the surface of the wave, so are different appliances necessary to register different disturbances of the luminiferous æther. A wave which is many metres long can only be perceived by its heating effects, or by electro-magnetic means which fail, when the scale is diminished so that the wave-length is reduced to a fraction of a millimetre. When the length is equal to about a thirteen-thousandth part of a centimetre, the physiological function of our retina begins to be effective and we can use optical apparatus, until the length is still further reduced to about half that value, when our eyes fail and we must have recourse to photographic means. There is still a gap between the shortest wave which has been produced electro-magnetically and the largest wave which has been measured thermally[1], but this gap is gradually being bridged over. While the " Rest-strahlen" of Rubens and Aschkinass have very substantially increased the range of measurement for heat radiations, electric waves have been reduced in length by Righi, Bose, Lampa and others.

The generation of electric waves on a large scale by its far reaching application in wireless telegraphy has spread in the public mind an almost awe-inspired sense of the wonders of science. To many persons wireless telegraphy appears to be the greatest of recent scientific discoveries, and in the public mind it seems more marvellous than the breaking up of atoms through their own internal instability. I do not wish to detract from the practical importance of the results achieved, or underrate the high merits of those, who by overcoming most serious practical difficulties, have furnished us with a means of communication which is likely to have an increasing importance. I insist, however, that as a scientific principle, the invention has taught us nothing of conspicuous novelty. In the first place, wireless telegraphy though historically arising out of the investigation of " Hertzian " waves, involves no scientific facts which were not known to Faraday. It

[1] Thermal effects accompany of course all disturbances of the medium and give us the only true measure of their intensity, the difficulty lies in the production of a quasi-homogeneous disturbance the wave-length of which can be ascertained.

depends on electro-magnetic action at a distance, independently of the time of transmission which may be instantaneous or not. If in the year 1860 a physicist had been asked, whether it were possible to transmit electro-magnetic signals to a distance, he would have answered that he actually does it every day in his laboratory, when he closes an electric circuit and deflects a galvanometer needle. If his questioner had continued to ask whether these signals could be transmitted beyond the range of his immediate surroundings, he would no doubt have replied that this depended on the scale of the experiment. He might, quite likely, in an incautious moment, have permitted himself to fix a distance, and to express his conviction, that no power was ever likely to be employed, and no receiving instrument of sufficient delicacy, ever likely to be invented, which would allow signals to be transmitted to more than a quarter of a mile; but I do not think he would ever have acknowledged that any limit of distance was fixed except by practical difficulties. Telegraphy without wires is older than that through wires, and was witnessed by the person who first saw bits of iron attracted, when a loadstone was brought near to them. The induction effects of strong currents rapidly changing in intensity are also examples of wireless telegraphy which were long known to affect instruments at considerable distances. The man who looks upon his "wire" as on a common-place phenomenon, while he considers his "Marconigram" an almost super-natural message, is not very far removed in scientific education from the old woman who hung a pair of boots on the telegraph wires, so that they should reach her son more quickly than if they were sent by parcel post. An eminent man of science, to whom I once expressed my astonishment that wireless telegraphy was considered to be such a novel and extraordinary achievement, illustrated this general attitude by the remark, that, if people had been accustomed to speak all their lives through speaking tubes they would be greatly surprised by the discovery that they could also speak without them. This puts the case exactly, the " tubeless speech " is the accurate counterpart of the "wireless telegram."

The experiments of Hertz repeated and extended by Lodge and others, rapidly secured the final triumph of the electro-magnetic theory, and models of the æther illustrating the new ideas were soon forthcoming. One essential point to be attended to was, that the æther should be incompressible, and a further specification which was not essential but generally accepted for the sake of simplicity, made the displacement of the æther consequent on an action of electric force linear and in the direction of that force. This carried with it the supposition, that an electric current through a conductor means the flow of an incompressible æther through that conductor. The mechanism of the formation of an electrostatic charge at the surface of the conductor would then be represented by a model in which an incompressible liquid is forced through a tube, which is closed at its ends by a flexible elastic membrane. The liquid would push out the membrane until the elastic reactions balanced the pressure of the fluid ; and near the end of the tube, there would be elastic strains and stresses in the partially yielding medium outside. These strains and stresses represented the electric strains and stresses. I have entered into this question because the mechanical representation of the mechanism of the electric current adopted at first as a picture of what might happen, soon became an article of dogma which delayed further progress.

If we only had to consider the passage of electricity through solid conductors, we might still be under the sway of the simple representation which I have sketched out; but neither conduction through electrolytes nor through gases could be made to fit into this simple scheme. Though neglected for a time, liquids and gases had ultimately to be taken into account. Faraday's laws of electrolysis had proved, that in the case of liquid conductors the passage of a certain quantity of electricity is always associated with the transference of a definite quantity of matter, and had suggested that each atom or radical which was set free by electrolysis, carries a definite quantity of electricity. Maxwell felt the difficulty of explaining Faraday's laws by means of his own views on conduction, and realised the importance of clearing up the apparent antagonism. The Chapter on Electrolysis in his *Electricity and Magnetism* forms very instructive reading. "Of all electrical phenomena," he writes, "electrolysis appears the most likely to furnish us with a real insight into the true nature of the electrical current, because we find currents of ordinary matter and currents of electricity forming part of the same phenomenon." After explaining Clausius' theory in detail, Maxwell continues: "But if we go on and assume that the molecules of the ions within the electrolyte are actually charged with certain definite quantities of electricity, positive and negative, so that the electrolytic current is simply a current of convection, we find this tempting hypothesis leads us into very difficult ground." Further on we find the following significant passage : " Suppose, however, we leap over this difficulty by simply asserting the fact of the constant value of molecular charge, and that we call this constant molecular charge, for convenience in description, *one molecule of electricity*. This phrase gross as it is, and out of harmony with the rest of this treatise, will enable us at least to state clearly what is known about electricity and to appreciate the outstanding difficulties."

How the " molecule of electricity " finally triumphed, and Maxwell's apostles who for a time persisted in repudiating it, had to bring their views into harmony with it, must now form the subject of our consideration. It is doubtful whether the phenomena of electrolysis alone would have been sufficient to lead to the present electron-theory, because it might have been possible to treat the quantitative relationship between electrolytic decomposition and current as a secondary phenomenon. The æther current in the crude view adopted by the early followers of Maxwell might pass through a liquid as through a solid conductor, and produce the observed results through the secondary action of electro-chemical effects. I imagine that several physicists were trying to develop some such idea; at any rate I did. (I have some recollection of explaining my views at a meeting of the Physical Society of London, but the paper was never published.) I soon convinced myself that the "molecule of electricity" could not be explained away, and turned my attention to the electric discharge through gases.

The history of the investigation of the passage of electricity through gaseous media, may serve to point several lessons. One of them is the truth of the statement, which has frequently been made, that the history of scientific discovery is to a great extent the history of scientific instruments and appliances. It is illustrated in our case by the marked advances which have accompanied each improvement in the construction of air pumps. The spectroscopic work of Plücker owed a good deal to "Geissler," whose pump brought down the pressure to a fraction of a millimetre. Later on, Crookes felt the want of better vacua, which again soon led to notable advances of our knowledge of gas discharges. Finally the requirements of glow lamps and Roentgen tubes made our demands still more severe, and again new discharge phenomena were brought to light.

The frame of mind with which the academic physicist looked upon investigations of the passage of electricity through gases, might be made the subject of instructive comment. The facts so far as they had been ascertained did not fit in with recognised views : hence they were ignored and students were warned off the subject. There was a feeling that perhaps in a century or so, the question might be attacked, but that in the meantime, it had better be left to be played with by cranks and visionaries. No criticism was more frequent at that time, than that of characterising as premature any new idea or fresh line of investigation in this direction ; as if any advance of science has ever been made which was not premature a fortnight before it was made. It is perfectly true, and just as much the effect as the cause of the attitude assumed by academic science, that a vast quantity of time and labour had been spent on the subject without making any material advance. There were a great many pretty experiments which seemed to defy all rational explanations, and were too complicated to teach us anything very definite. If we look back upon these experiments now, they may be used to point the moral that experiments conducted in what is sometimes considered to be the true philosophic spirit, where the investigator without any preconceived theories or notions simply wishes to classify facts, seldom lead to any valuable results.

Progress began when the subject was attacked with some definite object in view, either some theory however crude which had to be supported or some numerical connexion which had to be investigated.

The combat between rival theories centred to a great extent round the explanation of the phenomena, which were observed in partly exhausted tubes round the wire, which is connected to the negative pole of the electric battery or other source of electricity. Plücker had found during his spectroscopic experiments that as the gas was gradually exhausted, this wire (the kathode) becomes surrounded by a luminous glow, which expands as exhaustion proceeds (*N* Fig. 2). Hittorf, who had been associated with Plücker, found subsequently that bodies placed in this glow could cast a shadow; and Goldstein discovered further important facts, more especially with respect to the effects observed when two kathodes are placed side by side.

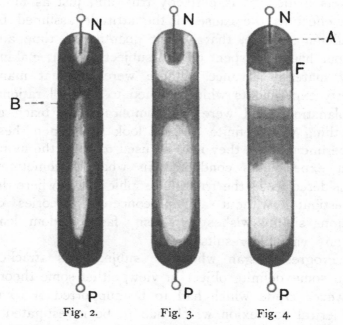

Fig. 2.   Fig. 3.   Fig. 4.

When the vacuum is sufficiently perfect, it is found that the glow separates off from the kathode, and leaves a dark space between the electrode and the luminosity, as shewn in Fig. 3. As exhaustion proceeds this dark space increases in width and ultimately extends to the boundary of the vessel. The transparent walls of the vessel, where they cut into the dark space (between *N* and *F*, Fig. 4), then become vividly fluorescent. It is when the tube is in this state, that a body placed in the dark space (e.g. at *A* Fig. 4) casts a shadow on the fluorescent portion of the wall.

Different investigators adopted different interpretations of the phenomena, which have just been described. The shadow effect suggested radiation as the cause of the glow, but radiation may be oscillatory or corpuscular in character. The natural inclination of those who remembered how a similar antagonism between two theories had, in the case of light, been decided in favour of oscillations, was to discard corpuscular theories as belonging to the middle ages. Hence many attempts at artificial explanations were made, while the simplest one was set aside. Goldstein's experiments, in which a second kathode placed parallel to the first was observed to repel the radiation coming from the first, seemed absolutely conclusive in favour of some theory of projected particles, but Goldstein himself took a different view, in spite of the fact that Helmholtz, in whose laboratory these experiments were made, urged him—as he subsequently assured me—to adopt the corpuscular hypothesis. Goldstein's experiments were begun about the year 1874, and soon afterwards Sir William Crookes, who had notably improved the means of obtaining high vacua in connexion with his radiometer experiments, began his work on kathode rays. His beautiful experiments soon attracted attention : they are too well known to need description here as we are mainly concerned with theoretical aspects. Crookes adopted the corpuscular view—a view which probably was first put forward by Varley in 1871—and by means of accumulated evidence of a most varied nature seemed to many of us to prove his case. Nevertheless a good deal of apathy was shewn, even in this country, with regard to the theoretical significance of the experiment, while in Germany the opposition to the corpuscular view was almost universal. The cause I believe to be this : although Maxwell's electrodynamic theory had not been generally accepted, the view that a current of electricity was only a flow of æther, appealed generally to the scientific world and was held almost universally. The most absurd consequences were sometimes drawn from this view, and Edlund's assertion that a perfect vacuum was a perfect conductor found many adherents even in England. Edlund's idea was, that a flow of æther could take place without resistance, where there is no matter, and that the high resistance of the most perfect vacua we can produce, is due to a resistance at the surface of the electrodes ; but this supposition ignores the fact that no electrical force can be transmitted through a perfect conductor however great the resistance at the surface may be. Adopting the view that a current of electricity simply means a flow of æther, it was tempting to attribute the effect observed under reduced pressures to secondary effects, accompanying longitudinal or other vibrations set up by the discharge. Attempts were made altogether to disconnect the luminous effects observed, from the primary effects of the discharge. Notably Heinrich Hertz conducted experiments which were intended to prove that kathode rays produced no magnetic effects, and therefore could not form part of the main process of conduction ; but these experiments, as I pointed out at the time, did not support the interpretation which Hertz gave them.

Before entering further into the discussion of the

nature of kathode rays, some other researches dealing with the discharge of electricity through partially exhausted vessels must be mentioned. A high place in the historical development of the subject is deservedly given to Hittorf, who first introduced accurate methods of measurement, more especially in the investigation of the relationship between current and electro-motive force. In order to obtain steady currents it was necessary to discard the induction coil, by means of which experiments had till then generally been conducted, and to replace it by primary cells, which were however, then also open to the suspicion of giving only intermittent discharges through gases. Gassiot was the first to put together a battery of primary cells having a sufficiently great electro-motive force (first 3620 zinc-water-copper elements, later 400 Grove cells) to send a current through a gas, but when he examined the image of the discharge in a rotating mirror he found it to be discontinuous; hence he concluded that the current was intermittent. The same result was obtained by Warren de la Rue, who with the help of a telephone satisfied himself that the current obtained from his chloride of silver battery was always discontinuous. Hittorf, however, after a careful investigation, was able to formulate the conditions under which an electric current can be obtained in a gas, which, so far as all available tests shew, is perfectly steady. In 1874[1], he experimented with 400 bichromate of potash cells, and in 1879[2] he described the construction of 1200 additional cells,

---

[1] *Poggendorff, Jubelband* (1874).
[2] *Wied. Ann.* VII. (1879).

giving him together with the previous set a total electro-motive force of over 3000 volts. It would lead me too far to describe the instrumental methods which were introduced by Hittorf, and generally adopted afterwards; but from the point of view of these lectures the following result is important. Hittorf shewed that, in the positive column of the discharge (*P, B* Fig. 2) so long as the pressure remains constant, the electric force driving the current along the tube is independent of the current. This result seemed quite paradoxical to those who regarded the current as a flow of an incompressible fluid, for when such an incompressible fluid flows through a tube, the difference of pressure at the ends, which corresponds to the electric force must necessarily increase, when the flow is increased. Nevertheless it will be seen presently, how simple the explanation of Hittorf's experiment becomes, according to our present view of the nature of electricity. Hittorf also obtained important results, by measuring the total electric force at the kathode. If the kathode is sufficiently large, the glow does not generally cover it, and he found that if the current be increased, while the glow expands and covers an additional area of the kathode the electric force does not alter. Only when the current has increased sufficiently to allow the glow to cover the whole of the kathode, is a further increase of the current accompanied by an increase of electric force. These researches were published in 1879.

The preceding account describes, I hope fairly and sufficiently, the state of the question, so far as I was acquainted with it, when I began experimenting with a view to obtain an insight into the mechanism of gas-discharges. I have already referred to the position taken up by Varley, Crookes and others, who contended that the kathode ray consisted of electrically charged projected particles, of which the velocity was due to electrostatic repulsion from the kathode, but the question how the particles of gas became charged had not been definitely raised. The most natural idea, and I believed the one generally held, was that the molecules of the gas took the charge by contact with the electrode.

While trying to bring Faraday's laws of electrolysis into harmony with Maxwell's theory, I became convinced that the most hopeful direction of research consisted in starting from the hypothesis that electricity was always concentrated in definite quantities, charging with definite amounts the carriers which conveyed the current. This was a logical and as events proved a correct step. In the case of electrolysis, it was further necessary to stipulate that the charges should never become detached from the atom, except at the surface of the electrode. It seemed natural and even obligatory to adhere to this restriction. The separate existence of a detached atom of electricity never occurred to me as possible, and if it had, and I had openly expressed such heterodox opinions, I should hardly have been considered a serious physicist, for the limits to allowable heterodoxy in science are soon reached. Looking back now with the advantage of the wisdom which comes after the event, I still think that in the first effort to explain gas discharges, one was forced to adopt the recognised explanation of electrolytic conduction, with as small a number of additional hypotheses as possible.

Hence it was necessary to retain the view that electricity is always attached to ponderable matter, so long as the facts could be brought into harmony with this view.

One fundamental distinction between the passage of electricity through electrolytes, and that through gases lies in the observation that an electro-motive force however small produces a current in a liquid, while in the case of gases a finite and sometimes a considerable electro-motive force is required to cause a discharge. But it must be remembered, that the behaviour of gases is really the more easy to understand, and that to explain the behaviour of liquids an additional and rather artificial hypothesis,—that of dissociation—had to be invented. The possibility of passing a current through a liquid by means of the smallest electric forces was originally a stumbling block in the theory of electrolytic conduction, but when the same theory was applied to gases this had been forgotten, and the more natural view was now objected to. A more real difficulty in the electrolytic view of gas discharges, lay in finding a reason for the opposite charges which it was necessary to ascribe to the two atoms forming the molecule of the ordinary gases; one of these had to be looked upon as being charged positively, while the other was negatively electrified. It is true that it was not necessary to assume that the polarity of the molecule was permanent, as it might only be established by the electric field, tending to drive the negative electricity in one direction and the positive in the other. Assuming the polarity to be established it became natural to suppose that with a certain critical value of the electric force, the molecule broke in two, and that the decomposition of a certain number of molecules then allowed the passage of electricity to take place by a process of diffusion. The view that electric conduction in gases is due to a diffusion of ions, similar to that which takes place in electrolytes, was first definitely proposed by Giese[1], in his investigation of the conductivity of flames. His work had escaped my attention at the time, but except in so far as the fundamental points of the theory are concerned, our work did not overlap, as I was mainly concerned with the discharge in partially exhausted vessels. In order to bring the phenomena which take place in the neighbourhood of the kathode within range of the theory, it was necessary to assume that the decomposition of the molecule took place near the kathode, whence the negatively charged particle was driven away with violence. The projected particle then formed the kathode ray. Such were the essential features of the theory I proposed in a paper which was presented to the Royal Society in 1884, and which, at the suggestion of Sir George Stokes, was chosen as the Bakerian lecture for that year. The view I had formed of the kathode glow seemed supported by experiments which appeared to shew that the most prominent features of that glow were absent in mercury vapour which is monatomic. Later, I convinced myself that an abnormally large dark space could form in mercury, and that its absence in my

---

[1] *Wied. Ann.* XVII. (1882).

first experiments was due to the narrowness of the tubes which were used.

I realised at an early stage that in order to demonstrate the correctness of the theory of ionic charges it was necessary to find a proof that the charge is a definite quantity, and that a crucial experiment could be devised by observing the magnetic deflexion of kathode rays. I pointed out that the experiments then available tended to support the view that the charge is indeed constant. As regards the positive discharge I shewed that Hittorf's laws were easily accounted for, by supposing that an increase of current may be produced by an increase in the number of particles which take part in the discharge, without change in their velocity. As it is the velocity which is determined by the electric force, an increase of the latter does not necessarily accompany an increase of current intensity.

Pursuing the subject I published a further paper in the year 1887, in which it was shewn, that a gas can be converted into a conductor by an independent discharge which is made to pass through it. Hittorf had already found that a column of gas through which a current passes, responds to small electromotive forces introduced at right angles, but his experiment left it undecided whether this transverse conductivity is due to peculiar conditions either of temperature or luminosity accompanying the primary discharge. In my own experiments, the conductivity was found to exist some distance away from the primary discharge. The experiments were explained by the breaking up of the neutral molecules in the primary discharge, the charged atoms acting as we would now say as ions capable of independent diffusion and therefore converting the whole mass of gas into a conductor of electricity. Similar experiments leading to the same conclusion were made by Arrhenius, and communicated by him to the Swedish Academy of Science on September 14th, 1887.

As the word "ion" is so frequently used in our discussion, it is well to make its meaning clear. Every carrier of electricity of not more than molecular dimensions is an ion, whether the carrier be an atom of electricity, or a charged molecule, or a portion of a molecule carrying a charge. The charge is measured by its external effects so that a molecule containing an equal number of positive and negative charges would, for our present purpose, be counted as uncharged. In the absence of any ionizing agent, the molecules of ordinary gases would be neutral, so far as we know. When through an external radiation or through any other process, the neutral molecule loses a charge, or splits up into charged portions, we say that the gas has become "ionized."

This ionic theory of gas discharges, while ignored in England, made good progress abroad; Arrhenius adopted it as well as Elster and Geitel, who were then investigating the electric behaviour of gases in the neighbourhood of incandescent electrodes. Warburg, whose important researches on the constancy of electric force at the kathode were made during this period, also expressed himself in general agreement with the views which I had expressed. In the meantime I was pursuing the subject experimentally, and was more especially trying to obtain evidence of the constancy of the charge carried by the ion, when I received a request once more to deliver the Bakerian lecture in 1890. It was intended that this lecture should give a general account of our knowledge on the conduction of electricity through gases. So far as my own experiments were concerned, the request came rather prematurely, as they were only in a preliminary state. My primary object in writing out the lecture, was to shew that the principal facts observed could be explained by the hypothesis that gases may be converted into conductors by ionization and that the charge of the ion is a fixed quantity. One of the great difficulties I found in discussing the subject with my colleagues was to convince them of the possibility of having a volume electrification, meaning an electrification spread through a volume instead of being confined to the surface of a conductor. This was generally denied in a dogmatic fashion on the ground of the incompressibility of the æther. In my lecture, I tried to shew that volume electrification must exist whenever a current passes through an electrolyte, which is not homogeneous in composition, and that even in solid conductors, which are not rigidly at the same temperature everywhere, volume electrification can only be avoided by introducing something analogous to hydrostatic pressure for which there is no experimental evidence. At any rate the possibility of volume electrification is easily recognised, if we adopt the hypothesis of an atomic charge, and a dogmatic denial could not dispose of that possibility.

At the head of my paper, I was able to place a sentence of far reaching importance taken from

Helmholtz's Faraday lecture: " If we accept the
hypothesis that the elementary substances are composed
of atoms, we cannot avoid concluding that electricity
also, positive as well as negative, is divided into
definite elementary portions, which behave like atoms
of electricity." As a personal recollection I may add
that at that period, I occasionally met Helmholtz
during the summer holiday which he used to spend
at Pontresina, and he frequently enquired after the
progress of my experiments. I consistently received
helpful encouragement from him, as I did from no
one else, more especially in the prosecution of the
investigation of the magnetic deflexion of kathode
rays, which he quite realised would yield the key of
the position.

If we assume that a particle having a mass m,
carries a charge e and moves with a velocity v, a
magnetic field at right angles to the direction of
motion will deflect the particle which then will no
longer move in a straight line but in a curve having
a radius r. If the intensity of the magnetic field be
M, the following relation must hold :—

$$\frac{e}{m} = \frac{v}{Mr}.$$

If we assume that the velocity is that due to the
electric force V we can calculate v in terms of V and
in that case

$$\frac{e}{m} = \frac{2V}{M^2r^2}.$$

The quantities on the right-hand side are all
measurable, hence we may determine the value e/m

and ascertain whether it is the same as that found by electrolysis. For nitrogen atoms we should expect the value to be about 2000, while the number actually obtained by the experiment was about three and a half millions[1]. If this number be accepted as correct, the conclusion I ought to have drawn from it is that either the quantity of electricity carried in the gas discharge is very much greater than that conveyed by the ions in electrolysis or that the mass of the carrier is very much smaller. Though perhaps I ought to have had more faith in my own experiment, there was certainly the possibility of a large error introduced by the assumption that the velocity of the particle had the full value which it would reach under the effect of the electric forces only. This meant neglecting the retardation due to mutual impacts which in my case (the pressure being about ·3 mm. of mercury) might have been considerable. Viscosity would ultimately reduce the velocity of the particle to that appertaining to ordinary gaseous translation, and if this ultimate velocity were substituted for v, the calculated value of e/m comes very near that which is found in electrolytes. Even now I am not convinced that the conclusion I drew from my experiment was not perfectly correct. I worked at pressures which were considerably higher than any at which the later experiments were conducted, and it seems to me to be very likely that in the rays which I investigated the electron had already attached

[1] In the published account of the Bakerian lecture the value $1\cdot1 \times 10^6$ is given; but an error in the adopted value of the magnetic field was subsequently discovered. The correction was made and further details about the experiments given in *Wied.* LXV. (1898).

itself to the molecule of matter. But even taking the experiment with all its imperfections, it was sufficient to prove that the charge carried was much greater than any that could have been received by the molecule through contact with the electrode—and this was pointed out. The above experiment did not attract the attention of physicists; at any rate four years later Sir Joseph J. Thomson in a paper published in the *Philosophical Magazine* wrote :—" I cannot find any quantitative experiment on the deflexion of these (kathode) rays by a magnet." The same paper contains the description of an attempt to measure the velocity of kathode rays, the result being found consistent with the supposition that e/m has the electrolytic value for hydrogen.

The next important advance was made by Sir Joseph J. Thomson, in a lecture delivered before the Royal Institution, and printed in full in the *Philosophical Magazine*, October, 1897, the essential portions having already appeared in the *Electrician* for May 21st, 1897. Perrin had shewn by direct experiment that the kathode ray carries a negative charge, but some objections which might have been raised against the form of the experiment were now removed. The ratio of the charge to the mass was determined by combining the magnetic deflexion of kathode rays with their total energy, measured by the effect they produced in raising the temperature of a thermo-junction. This ratio (e/m) was found to range between 3×10^7 and 10^7. The same ratio was determined by combining the magnetic deflexion with the deflexion in an electric field, acting at right angles to the rays. The value obtained by this method was $\cdot77 \times 10^7$. The new interpretation given to these experiments was summed up as follows : " Thus on this view we have in the kathode rays matter in a new state, a state in which the sub-division of matter is carried very much further than in the ordinary gaseous state : a state in which all matter—that is, matter derived from different sources such as hydrogen, oxygen, &c.—is of one and the same kind ; this matter being the substance from which all chemical elements are built up." On May 21 of the same year Kaufmann communicated a paper to *Wiedemann's Annalen* in which it was shewn that the magnetic deflexion of the kathode ray is, as required by the theory of projection, proportional to the square root of the potential fall. The value found for e/m was 10^7, but the author expresses a difficulty in understanding that this number is apparently independent of the nature of the gas and of the metal used as material of the kathode. In the following year Kaufmann in conjunction with Aschkinass shewed that the electrostatic repulsion effects of Goldstein are consistent with the corpuscular theory, and Kaufmann now appears fully to accept that theory, repeating his former measurements and obtaining the more accurate value of $1\cdot86 \times 10^7$.

In the interest of historical accuracy a lecture delivered by Professor E. Wiechert to the " Mathematisch-physikalisches Institut" of the University of Königsberg on January 7th, 1897, deserves to be mentioned. He describes experiments on the magnetic deflexion of kathode rays which did not go beyond the previous work of other observers, but he recognises

Atomic Charge

for the first time the smallness of the carrier; and though perhaps on insufficient grounds, identifies the carrier with the electron, and assigns a mass to it which is between the four hundredth and four thousandth part of the mass of the hydrogen atom. Two years subsequently[1], an important research was published by the same author. In it, he shewed how an experimental method, first attempted by Descoudres, could be made to give practical results in determining the velocity of the particles in the kathode ray. The method is of great originality, and the execution was difficult and well carried out. Once this velocity is determined, the ratio e/m can be measured, but though arriving at a valuable confirmation of previous results, Wiechert did not effect an improvement on the accuracy of the numerical constant to be measured.

In the experiments, which so far have been mentioned the ratio of the charge to the mass was measured, but the smallness of the mass could only be inferred if it be assumed that the charge of the carrier is the same as that we calculated from experiments with liquid electrolytes. It was therefore an important advance when Sir Joseph J. Thomson, in 1898, described experiments, which allowed the charge to be measured directly. He made use of a discovery of C. T. R. Wilson, that the negative ions can act as nuclei in the formation of water drops. If the air be ionized by Roentgen rays and a cloud be formed by a sudden chilling of the air, the small drops which make up the cloud may be observed to descend slowly, the rate of descent indicating their size. If the quantity

[1] *Wied. Ann.* LXIX. (1899).

of water vapour, which is condensed is known, the number of drops, and therefore the number of nuclei, is determined. On the other hand, if the ions formed are set in motion by electric forces the current may be measured, and this current must be equal to the product of the charge, the number of ions, and the velocity they require under the action of the given electric force. The last of these quantities may be determined independently, and therefore making use of the number of nuclei found by the experiment in the falling cloud, the amount of the charge can be calculated. In the current electro-magnetic system of measurement the charge found by Thomson was $7\cdot3 \times 10^{-10}$. Certain sources of error were not sufficiently taken account of in these earlier experiments, and H. A. Wilson, improving on the method, obtained a value rather less than half of that obtained by Thomson[1]. In order to compare the numerical value found with the charge of an ion in an electrolyte, it is necessary to introduce a quantity which is not accurately known, for electrolysis can only give us the value of the product of the charge into the number of molecules, which a cubic centimetre of hydrogen contains. If an estimate of this number be made by other means a substantial agreement is arrived at. No one could doubt, after the experiments which have been described, that the electric charges we are concerned with in the conduction through gases and through liquids are identical. Even then, however, the general body of physicists remained indifferent to the fundamental importance of the results arrived at

[1] The experiments of Rutherford and Geiger by a different and more accurate method give $4\cdot65 \times 10^{-10}$.

until the meeting of the British Association at Dover in 1899, when Sir J. J. Thomson described some further experiments, and explained more fully the conclusion to which they inevitably led. The experiments themselves only confirmed the previous results ; but while in the previous investigations the ratio e/m was measured in the kathode ray, and the charge e was measured on the ion formed by Roentgen rays, both quantities were now measured for the same ion, formed at the surface of a metal, when ultra violet radiation falls upon it. The experiments proved, beyond any possibility of doubt, the smallness of the mass of the electric carrier, which is set free by ultra violet radiation. The lecture in which the above experiments were described, was delivered to the British Association on the occasion of the visit of members of the French Association, which met concurrently at Boulogne[1]. It at once carried conviction, and though to those who had followed the gradual development of the subject, it only rendered more certain what previous experiments had already plainly indicated, the scientific world seemed suddenly to awake to the fact that their fundamental conceptions had been revolutionised. A new era of science begins at this point, and I cannot do better than conclude my lecture with the passage in which Sir Joseph J. Thomson summarised his results:

"I regard the atom as containing a large number of smaller bodies which I will call corpuscles ; these

[1] The title which originally was "On the existence of masses smaller than the atoms," but was changed when the paper was published in the *Philosophical Magazine*, XLVIII. p. 565 (1899) to "On the masses of ions in gases at low pressure."

corpuscles are equal to each other ; the mass of a corpuscle is the mass of the negative ion in a gas at low pressure, i.e. about 3×10^{-26} of a gramme. In the normal atom, this assemblage of corpuscles forms a system which is electrically neutral. Though the individual corpuscles behave like negative ions, yet when they are assembled in a neutral atom the negative effect is balanced by something which causes the space through which the corpuscles are spread to act as if it had a charge of positive electricity equal in amount to the sum of the negative charges on the corpuscles. Electrification of a gas I regard as due to the splitting up of some of the atoms of the gas, resulting in the detachment of a corpuscle from some of the atoms. The detached corpuscles behave like negative ions, each carrying a constant negative charge, which we shall call for brevity the unit charge ; while the part of the atom left behind behaves like a positive ion with the unit positive charge and a mass large compared with that of the negative ion. On this view, electrification essentially involves the splitting up of the atom, a part of the mass of the atom getting free and becoming detached from the original atom."

LECTURE III

During the first few days of January 1896, I was returning to Manchester after a short Christmas holiday, and on my way home called at the Laboratory for my letters. I opened a flat envelope containing photographs, which, without accompanying explanation, were unintelligible. Among them was one shewing the outline of a hand, with its bones clearly marked inside. I looked for a letter which might give the name of the sender and explain the photographs. There was none, but inside an insignificant wrapper I found a thin pamphlet entitled:—"Über eine neue Art von Strahlen"—(on a new kind of rays) by W. C. Roentgen. This was the first authentic news that reached England of a discovery made at the end of the year 1895, which both directly and indirectly gave a tremendous impulse to experimental science. Before I left the room, and although my family was waiting for me in the cold outside, I had read and re-read Roentgen's account, which concisely but convincingly described the experiments, by which he had with remarkable ability, investigated and determined the main properties of the new radiation. Briefly they are as follows:—the radiation sets out from the place where the kathode ray strikes an obstacle. In the original experiment this was the wall of the glass tube, but as subsequently found, a metal target acts equally well or better. The rays are more penetrating than any that were known at the time of the discovery, but are absorbed more or less by all substances. The radiation acts on a photographic plate or on a phosphorescent screen which becomes luminous under its influence. If a leather purse containing coins be held up between the source and the screen, the shadow of the purse is only faintly indicated while the shadows of the coins are very dark. Similarly, if a hand be placed in the path of the rays, the bones are clearly marked in the shadow, shewing that they absorb the radiation more strongly than the fleshy parts.

The discovery of the rays was partly accidental, which means that it was the consequence of one of those hints, which sometimes an investigator receives unexpectedly, and which according as they are taken, or neglected, make or mar a reputation. In this case a batch of photographic films lying near a discharge tube, which was set up in order to repeat Lenard's experiment on the transmission of kathode rays through aluminium leaves, was found to be fogged. It might have been merely a bad set of films, but the hint was taken, and within a week the principal properties of the new radiation were disclosed.

Roentgen proved that, unlike the rays of light, the new rays are not refracted when they pass obliquely from one medium to another, nor could he detect any interference effects : hence they seemed to be different in kind from rays of light.

The interest which the discovery roused in the scientific world and the sensation it created generally may be imagined, and there were few laboratories in which attempts were not immediately made to repeat the experiment. This was not altogether easy, because few institutions were then provided with the appliances necessary to obtain so perfect a vacuum as that required for the purpose, and also because the English lead glass is much less suitable than the soft German glass to excite and transmit the rays. Almost at once the possibility of practical applications attracted the public, and more especially the medical profession. In the diagnosis of complicated fractures of bones, or of the location of extraneous bodies, it was clear that a method of great utility was now available. To me this had an unfortunate consequence. My laboratory was inundated by medical men bringing patients, who were suspected of having needles in various parts of their bodies, and during one week I had to give the best part of three mornings to locating a needle in the foot of a ballet dancer, whose ailment had been diagnosed as bone disease. The discharge tubes had all to be prepared in the laboratory itself, and where a few seconds exposure is required now, half an hour had to be sacrificed owing to our ignorance of the best conditions for producing the rays. More difficult problems also arose, as when I had to travel to a small manufacturing town in the north of Lancashire, in order to locate a bullet in the skull of a poor dying woman who had been shot by her husband. My private assistant completely broke down under the strain and excitement of all this work, and the experiments on the magnetic deflexion of kathode rays on which I was then engaged were seriously interfered with by this interruption ; but I must return to my subject.

The absence of refraction and of interference had led Roentgen to reject the idea that the radiation was, like light, due to transverse vibrations of the æther. He therefore tentatively suggested that they might be longitudinal vibrations. He was possibly led to this suggestion by the belief that the kathode ray, according to the then fashionable theory, was connected with a wave-motion. Now in the theory of light there has always been this difficulty, that if the æther were an elastic body in the ordinary sense, a transverse wave entering from one medium to another ought to be partly converted into a longitudinal wave. These longitudinal waves had never been traced, and Roentgen may have thought that the new discovery might solve a difficulty in the older theory of light. Be this as it may, the suggestion never met with much favour and there is not much to be said for it. As regards the absence of interference and of refraction, I explained in a letter to *Nature*, which appeared a few weeks after the announcement of the discovery, why it is not necessarily antagonistic to the view, that the Roentgen rays are transverse vibrations of very short period. The reason I gave was, that interference depended on the regularity of vibration, which requires the predominance of waves of certain lengths over others, and in such an impulsive wave as is likely to be generated by the impact of a kathode ray, I pointed out that the necessary regularity would probably not exist. I also gave reasons, why the absence of refraction could be explained by a theory of refraction, which was then already in favour, the slower wave-velocities in transparent media being connected with the resonance of molecular vibration, responding to the periodic force of the passing wave.

Experiments on the absorption of the Roentgen radiation by different bodies give interesting results, of which at present only one need be mentioned. The rays emitted by different tubes, notably when they are in different stages of exhaustion, differ considerably in their penetrating power; hence we must conclude that the Roentgen rays are not "homogeneous" and differ from each other, much as the rays emitted by ordinary sources of light are found to differ. We must defer the discussion of the nature of the Roentgen ray until we can connect it with other radiations, but it is necessary to keep in mind from the beginning one marked distinction between a kathode ray and a Roentgen ray; the former is deflected by magnetic force while the latter is not. This distinction was insufficiently realised at first by some, who even went so far as to deny the novelty of Roentgen's discovery. Hertz and Lenard had already shewn—it was said—that kathode rays could traverse metallic bodies, and there was no reason why the effects observed by Roentgen might not have been produced by some of the kathode rays traversing the tube in which they were generated. I remember in this connexion an interesting discussion on board an excursion steamer on the occasion of the Kelvin Jubilee. A German Professor stood up for Lenard's priority in the discovery of the new radiation. The whole thing, he said, was in Lenard's mind, when he carried out his researches. "Ah," said Sir George Stokes with a characteristic smile, "Lenard may have had Roentgen rays in his own brain, but Roentgen got them into other people's bones." Nobody enjoyed the remark more, or subsequently repeated it oftener, than the eminent Professor who had called it forth.

It obviously suggested itself to every one acquainted with the subject, that the new radiation might prove an effective ionizer of a gas, and it is therefore not surprising that several observers discovered independently that this was indeed the case. This ionizing property gave us for the first time, a method for studying the properties of ionized gases in detail and with accuracy. The previous methods did not easily adapt themselves, partly because, as when ultra-violet radiation is used, the ionization only takes place at the illuminated surface and not throughout the mass of the gas, and partly because the ionizers shewed too many complicated irregularities. I had tried, for instance, with small success, to study the law of conduction of ionized gases in a vacuum, some distance away from the kathode, finding it difficult to obtain consistent results.

Very soon after Roentgen's discovery was announced, the matter was taken up in the Cavendish Laboratory, by Sir J. J. Thomson, Rutherford and others, and the main laws which regulated the effects were soon brought to light.

To get a clear idea of their meaning, let us begin by asking, what is an electric current? We need not go back to first principles and discuss whether it is a transference of æther or not, but may rest satisfied with what we can directly observe in every case in which a current passes. An electric current means the transference of electricity just as a current of water means the transference of water. If now we adopt the view that electricity is concentrated at some points in definite quantities, as has been explained in my second lecture, the motion of these points must constitute a current of electricity. These centres of electric force possess what in ordinary language we call mass and can act as "ions." The meaning of the word ion has been explained in the previous lecture, and we have been long familiar with the manner, in which an electric current is transmitted through an electrolyte by the convection of these charged carriers. If the positive ions move in one direction, there will be an electric current in that direction, and similarly a motion of the negative ions constitutes an electric current, but it is important to remember that as regards all external effects the motion of a negative ion to the left is identical with that of a positive ion to the right. Hence, if negative ions move—say—towards the west, this constitutes an electric current to the east. It appears therefore that the measure of the intensity of an electric current which depends merely on the quantity of electricity transmitted, does not completely specify the process of conduction ; because it leaves it undecided, whether positive ions move in one direction the negative ones being stationary, or negative ions move in the opposite direction, the positive ones being stationary, or whether both kinds take part in the current. What has been said will make it clear, I hope, that the current depends on the relative velocities of the negative and positive ions, as they move past each other, and that we have agreed to take the direction of the positive ions for the direction of the current.

The positive and negative ions may differ in other respects than that of the charge they carry. As a rule they have different masses. The electron which is the elementary constituent of electricity has, as we have seen in the last lecture, a very small mass, but positive ions so far as is known at present, have masses of the same magnitude as ordinary molecules of matter, and the negative ions which we find in most cases in ionized gases are not simple electrons but have masses which are not very much smaller than those of the positive carriers.

Let us now take a gas which is subjected to some ionizing radiation, such as that emanating from a Roentgen tube. The rays traversing the gas are absorbed to some extent, the absorbed energy being partly utilised in producing a definite number of ions in each second of time. Does this ionization then increase indefinitely? No, because an ionized gas does not, when left to itself remain ionized. There is a steady recombination of the positive and negative charges, and the gas tends to return to its original neutral condition. I must remind you here of the significant distinction between a liquid and a gas, which has already been dealt with in the previous lecture. In the liquid electrolyte, the ions are always present ready to follow the smallest electric force that may be applied, while in the case of a gas, ionization has to be set up before an electric current can pass, and no permanent ionization can be maintained without permanent ionizing cause. The rate at which the ions recombine depends on the nature of the gas, but there is not much difference in the case of the gases with which we are most familiar, one recombination taking place at atmospheric pressure in each second of time for each thousand positive and negative ions. The number of recombinations is proportional to the number of ions of each kind, so that if the number both of positive and of negative ions is doubled we should get four times as many recombinations in the same time. In the experiments on recombination great care must be taken to exclude the presence of dust, as small or liquid particles seem to attract the ions and thus accelerate their recombination. Similarly the presence of moisture helps the recombination.

Returning to the consideration of a gas which is subjected to a constant ionizing agent, we see that the ionization must reach a definite limit, when the number of ions produced each second equals that which disappears owing to recombinations. We obtain in this way a gas ionized to a certain definite amount, so long as the ionizing agent remains constant. When the conductivity of such a gas is studied, it is found that under the action of small electric forces, the electric current produced is proportional to the force. The positive ions will move to one side, the negative ones to the other, and the total number which give up their charge to the two electrodes in unit time gives the current conveyed by the gas. This law of proportionality between electro-motive force and current is the law known as "Ohm's law," which holds in solid and liquid conductors for all intensities of current which have hitherto been studied. But in the case

of gases it ceases to hold when the electro-motive force
is increased, as it is found that a limit is soon reached
when for a given ionizing agent, the current remains
the same whatever electro-motive force we apply.   It
is easy to understand why this should be.   Let us fix
our ideas by imagining two equal parallel
plates *A* and *B* (Fig. 5), connected respectively to the two poles of a battery *E*.   If
a Roentgen radiation of constant strength
be sent through the space separating the
plates, the air will be ionized and there
will be a definite number of ions (say 1000
of each kind) formed in each cubic centimetre of gas, in each second of time.

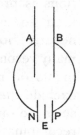

Fig. 5.

It is clear that in that case, one thousand charges represent the greatest quantity that can be communicated
to each electrode in one second, and the maximum
current that can pass through the gas is therefore equal
to 2000 units, multiplied by the total volume of gas,
measured in cubic centimetres, which is included
between the plates.   The maximum current will be
reached, when the electric force between the plates is
sufficiently strong to attract the ions, and remove them
as soon as they are formed.   The current is then called
a " saturation current."

The discovery of the Roentgen rays stimulated
scientific enquiry in many directions, but the most
remarkable results were obtained through a process
of reasoning ultimately found to be mistaken, which
may serve to shew that the ideas which induce us to
conduct an experiment are often of small importance
compared to the results which, if it be carried out in

a philosophic spirit, may be reached through it. Henri Becquerel, who following the tradition of his eminent father, had made himself thoroughly familiar with all phenomena of fluorescence and phosphorescence, had conceived the idea, that the emission of Roentgen rays might always accompany the excitation of phosphorescent light. He was led to this supposition by noting that in the original experiments of Roentgen, the rays were emitted from the spot where the kathode rays struck the glass rendering it phosphorescent. Having a variety of substances possessing the required property at his command, he tried to test his views, and after some unsuccessful trials placed two crystalline laminæ of a uranium salt on a photographic plate, which was wrapped in black paper, so that no ordinary light could fall on it; between one of the crystals and the photographic plates a piece of silver was interposed. Strong sunlight was then allowed to fall on the uranium salt, and after an exposure of several hours an action on the photographic plate could be detected, the outline of the crystal and the shadow of the piece of silver being distinctly marked. Here then, there seemed to be a complete justification of the view which had suggested the experiment, for a radiation had evidently penetrated the paper and reached the photographic film. It looked as if sunlight excited the uranium salt, just as the kathode ray excites the glass wall of a Roentgen tube. But now happened one of those chance occurrences, which while demolishing Becquerel's theories, placed him at the same time in the forefront of experimental discoverers. A plate was prepared by him in the way

6—2

described, but as the sun was only shining fitfully at the time, the experiment was discontinued, and after a short exposure the plate put in a drawer ready to be used again on a clear day. The next few days were cloudy, and the plate was developed without expectation of finding more than a feeble trace of the radiation : on the contrary, the shadow was marked with great distinctness. Becquerel was surprised but soon convinced himself that the sunlight had nothing to do with his previous results, but that his uranium salts always emitted radiations, which could penetrate paper and act on a photographic plate. It remained however to be seen, whether this was a permanent property of these substances, or only a temporary effect due to their having been previously exposed to some source of light. It might have been that the salt of uranium which had been lying in diffuse daylight for a considerable time, had absorbed a considerable amount of energy, which it was able subsequently to give out in the form of a penetrating radiation. If this were the case, we should expect the radiation to die out gradually though perhaps only slowly ; but experiment shewed no diminution of the effect with time. Hence Becquerel could claim with justice to have discovered a continuous radiation, involving an emission of energy by substances which till then had been considered to be unalterable and therefore containing a fixed amount of energy. The discovery of such a property naturally created a great sensation, and various observers examined a number of bodies, including all known elements, in order to see whether any of them emitted similar radiations.

All compounds of uranium shewed the effect to the same degree, if amounts were compared which contained equal weights of the metal, but thorium was the only other known element which was also found to be "radio-active." Madame Curie, who almost simultaneously with E. Schmidt, discovered this fact, made in conjunction with her husband, the significant observation that while all artificially prepared uranium compounds are equally active, pitchblende, which is a mineral consisting chiefly of uranium oxide, was weight for weight decidedly more powerful. To find the reason why, the various metals, which the mineral contains in small quantities, were separated chemically and examined. It was found that the extracted bismuth and barium were both strongly active, more especially the barium. Now ordinary barium shews no sign of any radio-activity, and the conclusion was therefore inevitable, that pitchblende contained some unknown element, having similar properties to barium, and separating out with that body in the chemical treatment to which it had been subjected. By a lengthy process often repeated, the new element was gradually separated from barium, and ultimately a product was obtained, which was about a million times more active than uranium. The name of "radium" was given to the element which was thus obtained, as chloride or bromide almost free from barium. Another radio-active element was found in the precipitate, which separated out with the bismuth, and was called "Polonium." Soon afterwards Debierne obtained a further active substance from pitchblende to which he gave the name of "Actinium." We had then apparently five elements possessing the new properties: Uranium, Thorium, Polonium, Radium, and Actinium. These elements, with the exception of Polonium give rise to other radio-active bodies, so that the total number known at present is much greater than five. Amongst them "Ionium" recently discovered by Boltwood is of special interest on account of its close relationship to Radium.

I now interrupt the historical account of these discoveries in order to discuss the entirely new aspect of the constitution of matter which they opened out to us. Clearness in perception of the fundamental facts and their theoretical bearing is here essential. We must begin by defining our terms, and making clear what is novel in the properties of radium and of its associates. We call a body radio-active if it permanently emits energy, which belongs to it in virtue of its chemical identity. The ordinary radiation of heat does not come within this definition, because though a hot body placed in a colder medium emits energy, it does not do so because it is iron or because it is copper, but because it is hotter than its surroundings. Similarly ordinary phosphorescent bodies are not radio-active, because the energy they emit is transient, and taken from a store which has previously been absorbed. The characteristic feature of the emission of energy from radio-active bodies is, that it is entirely independent of physical conditions. The three different kinds of "rays" which are emitted by these bodies are designated by the first three letters of the Greek alphabet, α, β, γ. The α ray is a projected particle which is positively charged and has a mass of about twice that of the atom of hydrogen, the β ray, on the other hand, while also corpuscular, has a much smaller mass, and is, in fact, identical with the kathode ray, being negatively charged and having a mass about the thousandth part of that of the hydrogen atom.   The γ radiation is identical with the Roentgen ray, and according to the generally accepted view, consists of a disturbance of the æther not unlike that of a group of extremely short waves. Our knowledge of the charge and mass of the particles which make up the rays, is mainly obtained from the deflexions observed when they are submitted to electric and magnetic forces.   If we consider the sign of the charges only, all we need observe is the direction in which a ray is bent, when placed in a magnetic field, and in that case we may represent the effect diagrammatically by Fig. 6, in which *R* represents a small

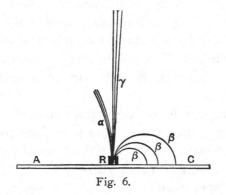

Fig. 6.

cup containing radium, and the magnetic force is supposed to act at right angles to the paper in such a way, that the north pole of a compass needle would be driven towards the back of the paper.   The diagram illustrates how the α rays bend very slightly to one side, while the γ rays are not deflected, and the β rays bend round to the other side in varying degrees according to the velocity with which they have been projected. The magnetic effect on the α rays is exaggerated in the figure, and is rather difficult to observe, so that it was thought at first that these rays carried no charge until Rutherford observed a slight bending, when the magnetic field was sufficiently increased.

The α rays are strongly absorbed by all substances on account of their small velocities ; the β rays are more penetrating, but not so much as the γ rays which, when of sufficient intensity, can be traced through a plate of lead over 20 cms. thick. The great difference in this absorbing power may be shewn by comparing the diminution of intensities of the different rays in the same metal such as aluminium, when it is found that 8 mms. of aluminium are required to reduce the γ rays sent out by a radio-active material to half their intensity, while half a millimetre is required for the β rays, and only the 2000th part of a millimetre for the α rays.

The magnitude of the spontaneous emission of energy which takes place from a radium compound is strikingly illustrated by the rise of temperature of the substance above that of its surroundings. This elevation of temperature which may amount to several degrees was first discovered by Curie and Laborde, and is accounted for by the absorption of the α rays in the radio-active body itself.

We now come to an important part of the lesson that may be learned by studying these phenomena.

An irregularity in the ionizing power of the radiation derived from thorium being traced to air currents, led Rutherford to the discovery of a gas, which is constantly being generated by thorium, and is responsible for part of its apparent radio-activity. This gas, which was called the "emanation," has all the properties of other gases; it diffuses at a definite rate which may be measured, and it solidifies when cooled down sufficiently. In every respect it behaves like an elementary body, but it disappears after a short life, so that in less than one minute after its first appearance only half the original quantity is left. No radio-active gas is given out by uranium, but radium also generates an emanation, and this has a longer life than that of the thorium emanation, so that half the original quantity is still left after about three days and twenty hours. The physical properties of the emanation of radium, owing to its longer life, can be more easily studied than those of thorium. Its spectrum has been measured, its boiling point fixed at − 64° C. for atmospheric pressure, and even its density in the solid state has been approximately determined.

Let us be clear about the facts. Emanations are gases which are continually being produced and continually vanish: both processes take place at a definite rate which hitherto has proved to be independent of all external conditions. A steady state may be reached, when the amount generated in each second of time (which depends on the quantity of the parent substance present), is equal to the amount that disappears which is proportional to the quantity of the disappearing substance.

If we take a solution of a radium salt, and extract all the emanation it contains, we must wait some days before a further quantity has been formed; but the process of formation continues as regularly as the process of self-destruction. What happens to the emanation as it disappears? Have we here really a destruction of matter? The answer to that question is suggested by the observation, that the walls of the vessel which has contained the emanation acquire radio-active properties, strong in proportion to the amount of emanation which has disappeared.

You are now in possession of the main facts, which led Professor Rutherford and Mr Soddy to a simple but far-reaching theory of radio-active phenomena. According to this theory, an atom or molecule is only radio-active while in the act of breaking up, and radio-active bodies are those which are unstable to the extent, that a certain definite fraction of their atoms break up every second of time. The breaking up of the atom is nearly always accompanied and probably caused, by the ejection of an α or of a β particle, or of both particles simultaneously; some of the charges are further accompanied by the γ radiation. The radio-active process is a consequence of the instability of the atom and consists of a succession of steps, one body being formed as the other disappears. When the ultimate product is inactive and does not give rise to any further emission of rays, the process ceases to come to our ken; for the quantities of matter at our command are, in most cases, so minute, that the radiation is the only property which affects our instruments. It will be noted that in calling a body "radio-active," we are guilty of the same inaccuracy as we should be in saying that the powder in a gun projects the bullet. The powder is the ultimate source of energy of the projection, but the projection itself is due to the disintegration of the powder, and not to the presence of it. Similarly a radium atom ceases to be radium by the very process which we call radio-activity.

The successive changes which radium undergoes, so far as is known at present, are illustrated in the diagram (Fig. 7):—

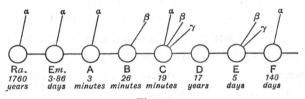

Fig. 7.

The radiation accompanying the change and the life of each intermediate substance is indicated in the figure, the life being measured by the time it takes for half the original quantity to disappear. Thus an atom of radium sending out an α particle becomes the atom of a gas (the emanation), and after the further emission of a similar particle, changes into a substance which is called "radium A," and has a life measured in the above standard by three minutes. The main β radiation only appears in the change from radium C to radium D^1. It will be noted that the change from radium D to radium E is not accompanied by any

---

[1] Some β rays of small penetrating power are already given out by radium B.

radiation. The existence of such rayless products cannot be directly proved but must be inferred from indirect indications.

If Rutherford's explanation be accepted as correct, as it universally is at the present time, it means that at least some of our chemical elements are in the act of breaking up. To say that radium is not an element does not dispose of the question, because if we deny to radium the claim of being a chemical element we must also do so to thorium and uranium and where shall we draw the line? Shall we revise our language and say that these metals are really compounds and not simple bodies? It would be a pity if we did, for there is no known distinction, except as regards their radio-activity, between uranium, thorium or radium, and other chemical elements. These bodies are certainly not chemical compounds in the common sense of the word. No doubt the name "atom" in its original meaning "uncut," and more especially in its traditional meaning "uncuttable," is no longer strictly applicable and this has given rise to critical comments on the unscientific use of scientific language. The late Lord Kelvin was a purist in such matters, and I was once present at a discussion in an English country-house, when he denounced, in forcible language to the guests, what he called the absurdity of speaking of the division of an undivisible body. He was interrupted by the son of the host, who was no respecter of authority, and turning round to the interested audience remarked, "There you see the disadvantages of knowing Greek." This remark derived some of its point from the discussion which then occupied the mind of Cambridge University regarding the advisability of keepin~ Greek as a compulsory subject in the entrance examination. It is impossible to modify our language continually, in order to make it keep pace with the advance of knowledge, and we must therefore be allowed to retain a word defining a group of certain entities, even if we have altered our ideas of the nature of these entities. The original name may etymologically be inappropriate, but as a label it retains its meaning. We shall continue to speak therefore of atoms just as we continue to speak of sunrise and sunset, which according to our present view are also incorrect expressions.

Let us return to the gradual transformation of the radium atom. If radium be left to itself for a considerable time, it must be accompanied by all its products of decay in such quantities, that for each atom of radium that disintegrates an atom of each of the products must also disappear. This will readily be understood if it be remembered, as already explained that an equilibrium state in the amount of the emanation is reached when one atom of the emanation that disappears is replaced by one atom generated. But each atom of emanation that is generated implies the disappearance of one atom of radium. When the same reasoning is applied to all bodies of the series, our statement will be seen to be correct. If all intermediate products be taken into account, it is found that four α particles are ejected for each atom of radium that disintegrates, and if we measure the whole number of α particles which are ejected by radium in its complete transformation, we obtain a measure of the number of radium atoms which break up in a given time. Such measurements and calculations have led to the result[1], that if we start with a certain quantity of radium, half of it will have disappeared after 1760 years and the millionth part only would remain after 30,000 years.

If this be true, how is it that there is any radium on the earth? Ought not all measurable quantities to have disappeared over and over again in geological times? The answer is obvious. Radium must continuously be forming as well as disappearing. Our endeavour, therefore, must be to discover its parent. Thorium and uranium suggest themselves as possible, both being present in pitchblende which is the main source of our radium, but various lines of argument point unmistakably to uranium and not to thorium as the ancestor of radium. It may be said, no doubt, that we are only pushing the difficulty further back, as we must now enquire into the parentage of uranium. We possess, at present, no information on that point, but it is to be noted that the rate of change of uranium is only about the one millionth part of that of radium, so that one gram of uranium would only lose one milligram in a million years. All science consists in pushing difficulties further back, and we may therefore be satisfied with having traced the ancestry of radium sufficiently far back, to include in our range of knowledge all radio-active effects which have been appreciable since the first solidification of the earth.

I must next speak of an important discovery suggested by the theory of disintegration. What

---

[1] Rutherford and Geiger, *Proc. Royal Soc.* LXXXI. p. 162 (1908).

becomes of the α particle ? This is a question which quite naturally presents itself. At its birth this particle has a positive charge, but it soon becomes electrically neutral either by losing its charge or by attaching to itself a negative electron from outside. For the purpose of identifying this neutral particle which must have a mass about twice that of the hydrogen molecule, Rutherford argued in this manner :—All final products of the radio-active change must collect and accumulate in quantity. The α particle is one of these products, and to identify it, we must look for an element which invariably accompanies radium or other bodies giving out α rays. Such an element is helium, which is invariably found in radio-active minerals. Now helium has a mass which indicates that it might itself be the α particle.

That helium is actually being produced during the decay of radium was subsequently confirmed by Sir Wm Ramsay working in conjunction with Soddy, and though the fact was at first received with some scepticism, it has now been proved beyond doubt. We may even measure the rate at which helium is generated from radium, the most recent determinations giving 158 cubic millimetres per year for each gram of radium.

The Hon. R. J. Strutt[1] having measured the amount of helium accumulated in rocks containing radio-active substances, by comparing the quantity of the gas present, with the amount of radium or uranium present, was able to draw some interesting and important conclusions. If no helium has escaped in the course

[1] *Proc. Royal Soc.* LXXXI. p. 272 (1908).

of time, the quantity present would give us a measure of the age of the rock since it was first deposited in a sufficiently coherent form to retain the gas. If some helium has escaped, we should still be able to form an estimate of the shortest time that the rock can have been in its present form. Unfortunately coprolites and phosphatised fossil bones are the only substances which allow an accurate investigation. As was to be expected the younger deposits do not give so large a quantity of accumulated helium as the older ones. The longest life-time found was that of a hæmatite overlying carboniferous limestone, which must have been in existence for at least 141 million years.

I have already given reasons why we should keep uranium and thorium in the list of bodies which we continue to call elements; the same reasons apply equally to radium and to all products of radio-activity, as no line can be drawn between uranium, radium, the emanations and the other more or less fugitive bodies which form the intermediate steps in the decay of radium. Some of these no doubt are very short-lived, but longevity has never been a decisive factor in a fundamental classification. We should therefore retain the word "element" but use it in connexion with the building up of compound bodies, in the sense in which bricks might be called the "elements" with which houses are built, irrespective of the question, whether bricks are themselves made up of smaller constituents.

What is the ultimate result of the radio-active process? The descendants of radium have been traced as far as radium F, which Rutherford has shewn to be polonium. This body on ejecting an a particle changes into an inactive substance, the identity of which is not yet quite certain, though Boltwood is probably right in thinking that it is lead, which thus can claim to be the immortal descendant of a family whose pedigree has been traced back to uranium.

With the facts and theories of radio-active change before us, we are able to consider the modifications of our former interpretations of physical phenomena, which recent discoveries suggest. Ever since Galileo, Huygens and Newton taught us the elements of dynamics we have considered inertia to be an inherent and fixed property of the centres of force, to which we have attached the idea of matter. Owing to the apparently rigorous proportionality between mass and gravitational force and the predominant influence of gravitation on all our doings in this world, we have acquired the habit of connecting mass and weight so intimately together that we can hardly separate one from the other. Nevertheless it is essential that we should do so, and place—as indeed Newton did—mass on an independent footing. The ratio of a force to the acceleration it produces may serve the purpose, or we may have recourse to the doctrine of energy. If a particle move with a velocity v and we take its kinetic energy, relative to the space in which it moves, in accordance with the common definition to be $\frac{1}{2}mv^2$, m is a factor which on the views of 25 years ago, belongs to the particle, as distinguished from the medium which surrounds it. Whether these views be correct or not, it is important to realise clearly, that energy cannot be expressed entirely by quantities, which merely depend on change of position (kinematical factors), but must involve something which the old idea of inertia was intended to cover. However much we may alter our views on the nature of inertia, our conception of natural phenomena must always, directly or indirectly, include this idea of inertia. The belief in the constancy of the factor m was justified by the experience, which taught us that in the visible motion of material bodies, energy seemed to be strictly proportional to the square of the velocity.

Without abandoning the old Newtonian dynamics, we should note, however, that the above simple derivation of mass from energy may lead us into error when applied to a system of connected bodies. Take, for instance, a heavy fly wheel (Fig. 8) of mass M and radius r, and let a string be wound round the central shaft of radius a, one end of the string hanging down with a mass m attached to the end. Neglecting the masses of the string and of the projecting portions of the shaft, the energy of motion may be expressed by $\frac{1}{2}\left(M\frac{r^2}{a^2}+m\right)v^2$, where v is the velocity of the mass m. If now, the fly wheel and the string were invisible, and we could only observe and measure the energy of the mass m, we might be misled into believing that its mass were $\left(M\frac{r^2}{a^2}+m\right)$. This fictitious mass, which may be great compared to M, is something which, partaking no doubt of the nature of a true mass, cannot be identified with any particular body or system of bodies, because

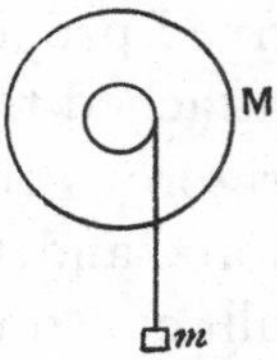

Fig. 8.

it involves the linear quantities r and a. If in the extreme case m becomes zero, the mass which would become identified with the particle is located entirely outside it. If further we imagine that, perhaps through centrifugal force, the radius r changes with the velocity, we should have constructed a model shewing how mass may depend on velocity.

Keeping the above in view, let us now estimate the energy associated with the motion of an electrified particle. Consider a sphere of radius a, uniformly charged with a quantity of electricity e. The expenditure of energy connected with the act of charging need not be taken into account, if for the sake of simplicity we assume the distribution of the charge on the sphere to remain constant during the motion. The moving charge excites a magnetic field in the surrounding space and, as first shewn by Sir J. J. Thomson, the total energy of the field is $e^2v^2/3a$. It follows that even though the sphere had no mass in the ordinary sense, it could not be set into motion without work being done upon it, and it would generally behave, as if it had a mass equal to $2e^2/3a$. It may be said that this mass is fictitious because we cannot locate it, but just as in the case of the invisible fly wheel of the above example, this does not abolish the idea of inertia, but only transfers that idea from the centre of force to the surrounding region.

Assuming the electron to be a charged sphere, we can calculate its radius by equating Thomson's expression for the mass to its value as determined from experiments on kathode rays. We obtain in this way a number approximately equal to 2×10^{-13} cms. As the molecular diameter is, approximately, 10^{-8} cms., an electron is something much smaller than a molecule.

The phenomena of radio-activity have proved, that the atom of such a body as uranium must contain a number of positive and negative electric charges and a certain part of the atomic mass must therefore reside in the electro-magnetic field immediately surrounding the atom. Whether that part is considerable, or perhaps even predominant, remains to be proved. But without waiting for such proof it is tempting as a plausible, though perhaps bold and risky hypothesis, to deny the existence of a Newtonian mass altogether, and to reduce everything to electric inertia. This leads us to a novel theory of matter, according to which an atom is entirely made up of negative electrons and positive charges. As the mass of an electron is about a thousand times smaller than the mass of the atom of hydrogen, it would follow that, assuming positive and negative charges to have equal masses, an atom of hydrogen is composed of about 500 electrons of each kind. But there is some doubt as to the proper way of treating positive electrons which have not hitherto been isolated. If the α particle consisted of a single positive charge, its large mass could only be accounted for, by assuming it to have a diameter a thousand times smaller than that of the negative electron, and the possibility of this seems to be excluded by the close relationship between the α particle and the atom of helium. It is reasonable, on the contrary, to conclude that the α particle consists of a large number of negative and positive charges, with an excess of one positive charge.

The great velocity with which electrons and α particles are ejected from radio-active substances shews that the ultimate components of the atom must be in a state of violent motion.

This suggests an atom to be something of the nature of a planetary system with the electric atoms revolving round each other as constituents. The question of the stability of such systems deserves, and has received, considerable attention, but no completely satisfactory solution has yet been found. A somewhat serious difficulty is presented by the dissipation of energy through electro-magnetic radiation, which accompanies all except a strictly uniform motion of electric charges. The electric inertia of a moving electron resides, as has been explained, in the electro-magnetic field surrounding the charge. If the velocity of the electron be constant in direction as well as magnitude, a steady state is reached when the magnetic field follows the charge in its motion, but as soon as either the velocity increases or the direction changes, a new field has to be established, and the change is effected by means of electro-magnetic radiations, propagated outwards from the moving charge with the velocity of light. The electron is thus called upon to supply a certain amount of energy which is dissipated into space, and this energy must have its equivalent in the work done on the electron itself, by a force resisting the change of motion. There is consequently always something in the nature of a retarding force whenever the motion of a system of electrons is such as to cause a change in the magnetic field. A single electron rotating round a fixed centre would quickly dissipate its energy into space by radiation, but if there were a continuous electrified ring involving no change in the magnetic field during rotation, there would be no such dissipation. A ring formed by separate corpuscles will approach the state of the continuous ring the more nearly the closer the electrons are together. Hence the dissipation of energy into space depends on the internal configurations. I must refer you to the original memoirs dealing with this subject for a further discussion of these and other matters, which have to be well weighed before an opinion on the all-electric origin of mass can be formed. The writings of Larmor, who first shewed the possibility of constructing an atom by means of moving electric charges, and of Lorentz, will shew that difficulties are not shirked, though they are not always solved.

While the scientific world is recovering from the shock of a great experimental discovery, there is generally a reaction, when explanations are showered upon it intended to demonstrate that so far from being surprising, the new discovery is exactly what ought to have been predicted. It was to be expected, therefore, that we should now be instructed how we might have foreseen radio-activity, if we had only adopted the electric theory of matter a little sooner. For if, as explained above, any motion in which the magnetic field does not remain strictly constant must gradually dissipate its energy, the rotating systems of electrons must sooner or later reach a state at which instability occurs. There may be the germ of truth in this and we may look forward to further developments of the idea, but many monsoons will pass over this country before the mechanism of an atom is laid bare to our understanding.

An important application of the dissipation of electric energy through radiation has been made by Sir J. J. Thomson in his theory of Roentgen rays. When an electron projected from the kathode meets with an obstacle so that its motion is strongly retarded, an electro-magnetic pulse spreads out from it into the surrounding space. This accounts for Roentgen's discovery in so natural a manner, that it has readily been accepted by men of science, and will not lightly be abandoned, though Professor Bragge has recently advocated, and given some grounds for the view, that the Roentgen radiation is, like the α and β ray, corpuscular in its nature and electrically neutral only, because two oppositely electrified particles are united in it.

Connected with the dissipation of electro-magnetic energy through radiation, because dependent like it on the finite velocity of light, is the change in the apparent mass of the electron when its velocity alters. The value I have given above for the mass of an electron only applies when its velocity is small compared with that of a wave of light, but we may conclude from the theory, that with increasing velocity the apparent mass increases until it becomes infinite, when the velocity of light is reached. This would mean that no velocity greater than the velocity of light is possible, which is an important deduction on which experimental confirmation would be welcome. We possess indeed experiments by Kaufmann and G. Bucherer which, in the case of electrons projected from a kathode, shew this increase of mass with velocity. A velocity of

more than nine-tenths of that of light was obtained,
and the apparent mass was found to be more than
three times as great as that for small velocities.
Experiments of this nature may furnish an answer
to the question whether the masses of electrons are
entirely electro-magnetic, or whether there is a remnant
of the old Newtonian mass. The answer, which at
present is perhaps not altogether decisive, is generally
taken as favouring the extreme view.

The state of plasticity and flux—a healthy state, in
my opinion—in which scientific thought of the present
age adapts itself to almost any novelty, is illustrated
by the complacency with which the most cherished
tenets of our fathers are being abandoned. Though
it was never an article of orthodox faith that chemical
elements were immutable and would not some day be
resolved into simpler· constituents, yet the conservation
of mass seemed to lie at the very foundation of creation.
But now-a-days the student finds little to disturb him,
perhaps too little, in the idea that mass changes with
velocity; and he does not always realise the full meaning
of the consequences which are involved. Does he
know, for instance, that the total mass of two electrons
placed side by side are not equal to the sum of their
two separate masses, but greater if they moved in the
same direction, and smaller when moving in opposite
directions. That electrons circulating inside the atom
never get sufficiently near for this change of mass to
become appreciable, does not seem to me an answer
to the difficulty, for it leaves out of account the
cumulative effect of the large number of electrons
concerned. Were it allowable to neglect this cumu-

lative effect, the inertia of an ordinary electrical circuit, would be proportional to its length, irrespective of its shape and the inertia of coil proportional to the number of its windings, while we know of course that it is proportional to the square of that number. The reason of the discrepancy becomes apparent when we consider that the electro-magnetic energy varies as the square of the electro-magnetic force, and that therefore, a thousand electrons arranged along an arc of a circle produce at its centre a field which per unit volume is a million times greater than that due to the single electron. Though we may therefore neglect the field beyond the immediate surroundings of a moving electron when dealing with it alone, we are no longer justified in doing so when a large number of electrons are concerned, even though their distances may be great compared with their dimensions.

If I draw attention to this change of mass, which according to the present theory must necessarily, to some extent, accompany all chemical combination, it is not in order to discredit that theory, but rather to point to a field of experimental research, which some day may lead us to a fresh experimental confirmation. So far however in spite of many careful efforts, mass as determined by weight has never shewn any tendency to be affected by chemical combination.

I have not yet finished my tale of the surprises sprung upon us by the theoretical pioneers, who at the present moment rival the experimental discoverers in boldness. One startling conclusion not quite the most daring, but running it pretty closely, arose out of an experiment made by the American man of science,

Michelson, and repeated later with much increased accuracy by him, jointly with E. W. Morley. The experiment was intended to clear up an obscure point in all our theories of light. If the æther is sufficiently solid to transmit torsional vibrations with a considerable velocity, how is it that matter can move freely through it. This difficulty frequently occupied the thought of Sir George Stokes, and previously Fresnel had tried to find out experimentally, whether, when a transparent body is in motion, it carries the æther filling the space between its molecules, with it, or whether that æther remains at rest. Fresnel's efforts remained undecisive because, according to our present ideas on the cause of refraction, the transmission of light through a moving body must be affected by the motion of the molecules, even though the æther remain at rest. The introduction of the electro-magnetic theory, getting rid of all idea of "solidity," seemed to lighten, and according to the views of Sir Joseph Larmor, altogether to remove this particular difficulty, if the æther be assumed to remain absolutely at rest. Michelson's experiment was intended to serve as a delicate optical test of the fixity of the æther by comparing the velocity of light in the direction in which the earth, as a whole, moves with the velocity at right angles to the motion. It is the orbital motion of the earth round the sun which comes into play, because it is much greater than that of the diurnal rotation, while the effects of a translation of the whole solar system through space may be eliminated by repeating the observations at different periods of the year.

The result of Michelson's experiment was the

velocity of light between two points on the earth's surface is the same whatever relation the direction of the line joining the points may have to the direction of the earth's orbital motion. At first sight this seems intelligible only on the view that the æther near the surface of the earth takes part in the orbital motion, but if this were true a large number of other difficulties would at once arise. The importance of the question makes it desirable to indicate the nature of the experiments, at any rate in outline. Let A and B (Fig. 9) be two points in the direction of the earth's motion, and C a point in a line at right angles to it and at the same distance from A as B. If a wave of light starting from A travels with velocity V, the distance l is traversed in time l/V, but by the time the wave reaches the point where B was, when the wave left A, that point has moved forward, owing to the bodily motion of the earth, which we imagine to be in the direction from A to B. The problem of calculating the time at which the wave reaches an object placed at B, moving with the earth, with a velocity v, is the same as that of finding the time when one body leaving A with a velocity V, overtakes another body leaving B with velocity v. This gives at once

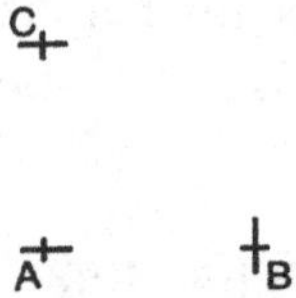

$$Vt = l + vt$$

and hence t is equal to $(l/V - v)$, and the actual length of the path traversed by the light in passing from A to B instead of being l is now $lV/(V - v)$. If there be a mirror at B which returns the light to A, we find

Fig. 9.

similarly that the path traversed in the return passage is $lV/V+v$, so that the total path, including the two journeys, is :

$$lV\left(\frac{1}{V+v}+\frac{1}{V-v}\right)=\frac{2lV^2}{V^2-v^2}.$$

If the fraction v/V be sufficiently small to allow its fourth power being neglected, the expression for the total path may be taken as equal to

$$2l\,(V^2+v^2)/V^2.$$

On the other hand, if a ray of light passes from A to a mirror at C and is reflected back to A, the path is found to be equal to

$$2l\,(V^2+\tfrac{1}{2}v^2)/V^2.$$

The difference in length between the paths is lv^2/V^2, and it is this difference in length that Michelson set himself to observe. The value of v/V is about one divided by ten thousand, so that the experiment, to be successful, requires the measurement of the hundred millionth part of the distance l; the resources of optics are equal to this task, thanks in great part to Michelson himself. It is of course impossible to adjust the distances AB and AC to absolute equality, but this does not matter provided that the whole apparatus can be rotated through a right angle, so that either length can be placed in the direction coinciding with that of the velocity of the earth. The conclusion finally arrived at by Michelson and Morley was, that so far as their experiments could decide, the difference in path, calculated on the hypothesis of a quiescent æther did not exist; but a quiescent æther is one of the few necessities of modern physics. How are we to meet this difficulty ? Mathematicians are always sufficiently resourceful to cope with any problem set to them by the experimentalist, and as Poincaré likes to tell us, we can always find a new hypothesis to fit a new fact. The hypothesis in the present instance was supplied independently by FitzGerald and Lorentz. The mathematical reasoning on which the above expression for the difference in optical path of the rays going respectively with the earth's motion and at right angles to it, is based, assumes that the length l remains fixed as the apparatus is rotated through a right angle. It seems possible at any rate that the dimensions of what we call a rigid body may alter according to the direction of its motion through space. If a contraction of appropriate amount took place along all lines in a moving body parallel to the direction of motion, Michelson's negative result could be explained. Further, Lorentz and Larmor have given grounds for believing that if molecular forces have ultimately an electric origin, such a contraction ought to take place. The matter did not rest there, mainly perhaps because the explanation was felt by some to be a little artificial. Looking at the experiment in a broad way it shews, that at any rate with our present methods of observation, we are unable to detect any relative motion between matter and æther, but only relative motion between different portions of matter. Let us take the point of view that this is not a consequential effect depending on a peculiar influence of velocity on molecular forces, such as is required by FitzGerald's explanation, but a law of nature holding universally. Einstein, in a paper of great interest and power, has developed this idea, calling his imagined law " The principle of relativity," because it stipulates—*a priori*—that only relative motion between material bodies can be detected. It is impossible for me to discuss in detail the reasoning by which this principle is justified, and an account without explanations of its consequences would lay me open to the charge that I was playing with your credulity. Suffice, therefore, it to say that strict adherers to the principle cannot admit the existence of an æther, and yet may speak of the transmission of light through space with a definite velocity. They must further accept, as a consequence of their dogma, that identical clocks placed on two bodies moving with different velocities have different rates of going and that, even on the same body, identical clocks indicate different times, when the line joining their positions lies in the direction of motion. The motion must be determined relative to another body, which is supposed to be at rest, and a clock placed on that body must serve as the ultimate standard of time. The theory appears to have an extraordinary power of fascinating mathematicians, and it will certainly take its place in any critical examination of our scientific beliefs; but we must not let the simplicity of the assumption underlying the principle hide the very slender experimental basis on which it rests at present, and more especially not lose sight of the fact, that it goes much beyond what is proved by Michelson's experiment. In that experiment, the source of light and the mirrors which reflected the light were all connected together by rigid bodies, and their distances depended therefore on the intensity of molecular forces. Einstein's generalisation assumes that the result of the experiment would still be the same, if performed in a free space with the source of light and mirrors disconnected from each other but endowed with a common velocity. This is a considerable and, perhaps, not quite justifiable generalisation. I am well aware that Bucherer's experiments with kathode rays are taken to confirm the validity of Einstein's principle, but if we say that they are not inconsistent with it, we should probably go as far as is justifiable.

In bringing the principle of relativity to your notice I have characterised it as one of the most startling developments of recent science, and you may ask why I have qualified the superlative ; have I reserved some even more surprising fancy of the scientific imagination ? That is indeed the case, though it may be difficult to place modern philosophic speculations in order of revolutionary merit. When in the presidential address to the Physical Section of the British Association at Edinburgh, I alluded to certain views on the transmission of energy which had been advocated, I stated that they could only be true if energy had, like matter, an atomic constitution. I thought I had thus finally disposed of the matter by a *reductio ad absurdum*, but now-a-days such atomic constitution is openly advocated. It all arose out of the theoretical investigations intended to account for the energy radiated by a hot and ideally black body, and more especially the distribution of that energy among the waves of different frequencies. The law of partition of energy which holds in this case is correctly represented by a formula which we owe to Max Planck, who deduced it theoretically by assuming that the radiation takes place in such a way that a molecule can only radiate either a certain definite minimum quantity or some multiple thereof. This means that as far as this type of energy is concerned, it behaves as if it were made up of a number of finite bits of equal value. As in the case of Michelson's experiment, this result has been generalised and the hypothesis has been formulated that all energy is made up of finite bits. That such a hypothesis should be advocated by men whose opinion deserves the most serious consideration, shews the restless turmoil which agitates the scientific thought of the present day. During the last century we felt sure that we were building our scientific edifice on a secure basis; to-day many have become suspicious of the soundness of its foundations, while some are already digging for new ones or tinkering with the old; there are still however a few optimists left who try to go on building upon the old structure in the perhaps mistaken belief that it still stands on firm ground, and will remain standing just as in the material world we find some edifices sufficiently strong to survive the ravages of an earthquake.

I confess to a feeling of relief in stepping now from the steep and slippery slopes of universal theories unto more modest but safer ground. Whatever may be the ultimate constitution of matter, there is no doubt that electrons are attached to the molecules and are responsible for the radiations of light and heat which they emit. It has already been briefly alluded to in the first lecture that Zeeman discovered a change in the waves emitted by an incandescent body, when that body is under the influence of a strong magnet. From the point of view of the electron theory this is easily explained, because the origin of the wave of light is an alternating motion of an electrified particle, which must obey the well-known laws of electro-magnetic force. Indeed, the discovery of Zeeman was incomplete until Lorentz applying theory pointed out to him that he ought to look for circular and elliptic polarisation. The magnitude of the effect cannot easily be predicted by theory, as it depends on the mutual interaction of the electrons which take part in the emission of light. In the simplest possible case, observation and theory combined gives us the relation of the quantity of electricity concerned, to the apparent mass of the vibratory system, and as this for some of the spectroscopic lines is found to be nearly the same as the ratio of quantity to mass deduced from experiments on kathode rays, we must conclude that in the simplest cases, the vibrating system consists of an electron which can vibrate independently of others.

It is not possible within the limits of this lecture to give an account of the various applications of the electron theory, which has been found capable of solving many difficulties, but some reference must be made to the work of Drude, which involves a new idea, and gives a rational explanation of the parallelism between electric and thermal conductivities. Drude imagines a number of free electrons within a conductor, which acquire a velocity in virtue of the internal thermal energy of the body. The average energy of these electrons should be equal to the average energy of a gaseous molecule at the same temperature and this is known. Heat conduction, according to Drude, is not due to the interchange of energy from atom to atom, as previously believed, but takes place through the intermediate action of these electrons. The reflexion of light at the surface of metals was also examined by Drude, and an appropriate explanation of the facts given by taking account both of the free electrons and of others attached to the molecules but free to vibrate. I shewed subsequently how an estimate may be formed from optical phenomena of the number of electrons which are capable of free motion in the metals, their number being from two to four or five for each molecule of the substance.

I have during these lectures contrasted on several occasions the former tendency to base our theoretical explanations of natural phenomena on definite models which we can visualise and even construct, with the modern spirit which is satisfied with a mathematical formula, and symbols which frequently have no strictly definable meaning. I ought to explain the distinction between the two points of view which represent two attitudes of mind, and I can do so most shortly by referring to the history of the electro-dynamic theory of light, the main landmarks of which I have already pointed out in the second lecture. The undulatory theory—as it left the hands of Thomas Young, Fresnel and Stokes—was based on the idea that the æther possessed the properties of an elastic solid. Maxwell's medium being quite different in its behaviour, its author at first considered it to be necessary to justify the possibility of its existence, by showing how, by means of fly wheels and a peculiar cellular construction, we might produce a composite body having the required properties. Although later Maxwell laid no further stress on the ultimate construction of the medium, his ideas remained definite and to him the displacements which constituted the motion of light possessed a concrete reality. In estimating the importance of the support which Maxwell's views have received from experiment, we must distinguish between the fundamental assumptions on which Maxwell based his investigations and the mathematical formulæ which were the outcome of these investigations. It is clearly the mathematical formulæ only which are confirmed and the same formulæ might have been derived from quite different premises. It has always been necessary, as a second step of a great discovery, to clear away the immaterial portions which are almost invariable accessories of the first pioneer work, and Heinrich Hertz, who besides being an experimental investigator was a philosopher of great perspicacity, performed this part of the work very thoroughly. The mathematical formula instead of being the result embodying concrete ideas, now became the only thing which really mattered. To use an acute and celebrated expression of Gustav Kirchhoff, it is the object of science to *describe* natural phenomena, not to *explain* them. When we have expressed by an equation the correct relationship between different natural phenomena we have gone as far as we safely can, and if we go beyond we are entering on purely speculative ground. I have nothing to say against this as a philosophic doctrine, and I shall adopt it myself when lying on my death-bed, if I have then sufficient strength to philosophise on the limitations of our intellect. But while I accept the point of view as a correct death-bed doctrine, I believe it to be fatal to a healthy development of science. Granting the impossibility of penetrating beyond the most superficial layers of observed phenomena, I would put the distinction between the two attitudes of mind in this way : One glorifies our ignorance, while the other accepts it as a regrettable necessity. The practical impediment to the progress of physics, of what may reluctantly be admitted as correct metaphysics, is both real and substantial and might be illustrated almost from any recent volume of scientific periodicals. Everyone who has ever tried to add his mite to advancing knowledge must know that vagueness of ideas is his greatest stumbling-block. But this vagueness which used to be recognised as our great enemy is now being enshrined as an idol to be worshipped. We may never know what constitutes atoms or what is the real structure of the æther, why trouble therefore, it is said, to find out more about them. Is it not safer, on the contrary, to confine ourselves to a general talk on entropy, luminiferous vectors and undefined symbols expressing vaguely certain physical relationships? What really lies at the bottom of the great fascination which these new doctrines exert on the present generation is sheer cowardice : the fear of having its errors brought home to it. As one who believes that metaphysics is a study apart from physics, not to be mixed up with it, and who considers that the main object of the physicist is to add to our knowledge, without troubling himself much as to how that knowledge may ultimately be interpreted, I must warn you against the temptation of sheltering yourself behind an illusive rampart of safety. We all prefer being right to being wrong, but it is better to be wrong than to be neither right nor wrong.

LECTURE IV

DIFFICULTIES cease to trouble us either when they are surmounted, or when we have become accustomed to them. As soon as we observe a rare phenomenon, our brain starts working to find some plausible explanation, while matters of daily occurrence, which may be the most puzzling of all, are treated with indifference. Every one, for instance, wants to know why a comet has a tail, yet how few ever trouble to think why the sky is blue or the sunset red ; and many a man wants to have wireless telegraphy explained to him, though he does not know—or care—how the ordinary telegraph works. This tendency affects not only those who take only a spasmodic interest in science, but perhaps even to a higher degree the professional investigator. For every hundred learned men who are deeply interested in an experiment which can only be shewn with the most delicate and costly apparatus, perhaps one troubles to think why the earth is a magnet or charged with negative electricity, and as regards the mystery of gravitation, we have long given it up as a hopeless subject to worry about. It is not that we should not like to know the causes of all these things, but we are so accustomed to meet the questions which are daily before us, that they cease to assert that spell of curiosity which is the necessary antecedent to their being successfully attacked. I propose to-day to bring to your notice a few subjects which have suffered on account of the contempt,— neglect would perhaps be a more polite word,— which is proverbially bred by familiarity.

The progress of knowledge which I described in the three previous lectures was made by means of what is called the experimental method of investigation. Here we fix our attention on some, perhaps, casual observation, and by varying artificially the conditions we subject nature, as it were, to a severe cross-examination, which, if it prove successful, may divulge some deeply hidden secret; but when, for instance, we wish to investigate the connexion of sun-spots with terrestrial magnetism, we are unable to affect the phenomena to be investigated, and are reduced to the purely observational method. We can do nothing but sit, watch, and note carefully, and be satisfied with basing our reasoning on comparatively few facts, because the work of an individual is limited to the time of his life. Yet this patient observation of facts which happen without our interference, often carries us further than what can be achieved by experiment on account of the larger scale in which the phenomena present themselves. In our laboratories we can reach and measure temperatures up to about 3000° C., but substances near the luminous surface of the sun are at a temperature which is as high as 7000° C. and probably higher. Here nature gives us the opportunity of extending experimental methods.

Similarly, we are able, by experiment, to subject matter to a pressure of a few thousand atmospheres, but this is as nothing compared with the pressures which must exist in the interior of the earth. Here observation allows us to supplement our knowledge of the properties of matter at high pressures, because whenever an earthquake of sufficient intensity occurs, waves pass through the interior of the earth and their rate of propagation indicates the rigidity and compressibility of its inner core.

The question of scale is often of paramount importance, and more especially in questions which affect our atmosphere. Though we believe we know something about the formation of clouds, we should get to know vastly more, if we could arrange experiments on a scale sufficiently large, to form an artificial cloud under conditions similar to those which hold in the actual case. But scale is also important as regards time, and the extent to which our scientific activity and our scientific judgment are affected by the duration of our lives has never been sufficiently recognised. We are by nature individualists and consequently wish to see the results of our own work. If we can substantially improve an investigation by an additional week's labour, most of us would do so, but if the same improvement would require ten more years, the policy of waiting would not commend itself to the same extent. Similarly, when we have followed the course of an experiment for a few months we are proud of our patience, and few of us only would undertake one extending over several years. When centuries are required, we give up the attempt; yet how grateful should we not be, if the old Egyptians, six or seven thousand years ago, had sealed up in air-tight vessels different substances, all carefully compared as regards their weight, so that we could now judge whether in the interval they had preserved the same relationship of mass. Our ancestors have done nothing for us, should we, having felt the want, not do something for our descendants?

I have spoken of the scale of time as affecting our scientific judgment, and perhaps I ought to explain my meaning. If the sun shines on three successive Fridays with cloudy weather in between, we should laugh at any man bold enough to predict on the strength of this observation that the sun will always shine on Fridays: yet supposing that in the year at which sun-spots are at their maximum, which occurs at intervals of about eleven years, the monsoons were exceptionally large or exceptionally small on three successive occasions, how many of us would resist the temptation to believe that there is a true connexion between sun-spots and the weather? What is the difference in the two cases? To my mind it is only the relationship of the period to the duration of our own lives. In both cases, the mathematical probability of an accidental coincidence is the same, but with the weekly occurrence the true nature of the coincidence is bound to be revealed within a short time, and the risk of being wrong is more deeply impressed on our minds, because a few weeks would be sufficient to find us out. In the case of sun-spots on the other hand, a decisive test can only be obtained after eleven, and possibly twenty-two or thirty-three years; and the length of time which must lapse before we can be shewn to be wrong in asserting a true connexion, increases the confidence with which we assert it. I am far from intending to throw ridicule on those who in attempting to relate definite long period phenomena base their reasoning on comparatively few data. We are bound to do so. The progress of science is achieved by individual reasoning, which cannot be transferred from one brain to another; and the average span of a life-time must therefore be an important factor in the rate at which science progresses. This is a tempting subject to dilate upon, but I must refrain from further moralising.

The science of Terrestrial Magnetism began with Gilbert, who taught us that the earth as a whole behaved as if it were a magnet. If you imagine a large sphere of magnetised steel, and investigate the magnetic field in its neighbourhood by means of a small compass needle placed at different points on its surface, this needle would, roughly speaking, point towards the magnetic poles of the sphere. We call a "north" pole, a magnetic pole, which tends towards a point not far removed from the north geographical pole of the earth; but, as you know, the north pole of one magnet attracts the south and repels the north pole of another magnet. If therefore the "north" pole of a compass needle is attracted towards the north geographical pole of the earth, this must mean that the earth's magnetic pole which is situated in the northern hemisphere, must be of the same kind as the pole of our compass needle which we call the "south" pole. To avoid the confusion which may arise from forgetting that geographical north means magnetic south, Gilbert took the bold step of calling that end of a compass which points north, the "south" pole of the magnet, and British men of science followed that practice until Faraday introduced the term "north seeking" pole, and this has been gradually replaced by "north" pole, thus bringing our nomenclature into accordance with continental practice. Unfortunately at the same time that this change took place in England the opposite change was made in France, so that now Gilbert's nomenclature is very generally adopted in that country, and uniformity has again been disturbed.

The earth possesses to a great extent that simple form of magnetisation, which is technically called "uniform"; consequently the forces on the surface of the earth are nearly the same as those which would hold, if there were a small magnet at the centre of the earth possessing one north and one south pole. The lines of force, which at each point indicate the direction in which a magnetic pole would be driven, are shewn in Fig. 10, the arrow-head pointing in the direction of the force acting on the north pole. We have no means of knowing how they run inside the earth, but outside and near its surface, the agreement of the actual magnetic forces with those of a small magnet at the centre holds approximately, though not perfectly. The line joining the north and south poles of our small imaginary magnet is called the magnetic axis and is shewn in the figure. If a small compass needle, pivoted so as to be able to turn freely in all directions, were placed at the places where the magnetic axis cuts the surface of the earth, it would point vertically downwards, and these places would then be what we call the magnetic poles of the earth. From the scientific point of view these magnetic poles defined as being the spots where a magnetic needle points vertically downwards, are of no importance, because as is easy to see, their position may greatly

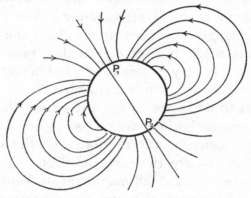

Fig. 10.

be affected by magnetic rocks, placed accidentally near P_1 or P_2. What is of paramount importance in describing the magnetic state of the earth is the direction of the magnetic axis, which may be defined in a strictly scientific manner, independently of the position of the poles.

This magnetic axis is observed to shift slowly, and our observations, so far as they reach, are consistent with a slow turning of the magnetic round the geographical axis in the direction from east to west. The change of the magnetic forces accompanying the shift of the axis is called the secular variation of terrestrial magnetism. Its study forms one of the

most interesting branches of the subject and one of
the most difficult, because accurate measurements at a
sufficient number of stations only date back a century.
The magnetic poles must shift with the magnetic axis,
but they probably do not keep pace with it in a regular
manner. If indeed the actual position of the poles
is affected by magnetic rocks, we can imagine a
considerable shift of the magnetic axis to take place
without a corresponding change of the magnetic pole,
which, when the axis is displaced sufficiently, may
then follow with a rush. Hence we should attach
but little importance to the exact determination of
the magnetic poles in arctic and antarctic expeditions.
What is important to the general theory of terrestrial
magnetism is the accurate determination of magnetic
elements in the region of the poles without undue
regard to one particular point.

Our magnetic observations are practically confined
to the surface of the earth, but it is remarkable how
much information we may obtain from these observations, by applying theoretical considerations derived
from a general knowledge of magnetic and electro-
magnetic forces. The general principle of the conservation of energy which holds universally, allows us
to conclude that when a small magnet is shifted from
one place to another, the work done can only depend
on the initial and final positions, but not on the path
taken during the displacement. When the magnet is
moved in a closed path so that the initial and final
positions are coincident, the total work must be zero.
Can we apply this principle also to the displacement
of a single pole? So far as the theory of energy is

concerned we must answer "No," because it is not allowable to go beyond what can be tested by experiment, and it is not possible to divide a magnet so that its two poles can be moved separately. But in the general analysis of magnetic forces, we resolve the total effect on finite magnets into elementary forces, between two poles, or between one pole and an electric current. The complete agreement between the derived and observed forces, justifies us in using these elementary forces as stepping stones. We are thus able to assert, that the work done in allowing a pole to describe a closed circuit is zero, when the magnetic forces are all derived from magnets, but not necessarily so when they are due to electric currents. If *AB* (Fig. 11) represents a portion of an electric current, and a magnetic pole at *P*

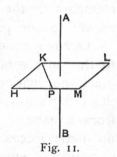

Fig. 11.

be moved round the current in a closed path (e.g. *PHKLMP* or *PKLMP*) until it is brought back to *P*, the total work done is not zero, but a definite quantity proportional to the total electric current which has been enclosed by the path. If the path (e.g. *PHKP*) does not include the current, the work is zero. This is important, because it allows us to decide by experiments made on the surface of the earth, whether electric currents of sufficient intensity traverse its surface. If we take the observations now available and apply the test to a circle of latitude, we should find indeed that these observations can only be reconciled, if we admit a substantial vertical electric current. On the other hand, when we apply the test to regions of the earth, which have been surveyed with great care, such as the British Isles (Rücker and Thorpe), we find that they do not indicate the existence of such vertical currents. There is therefore a strong suspicion that in the former case, the observations extending over regions including the ocean basins are not sufficiently accurate to allow any certain conclusions to be drawn, and this suspicion is strengthened by other lines of reasoning based on the direct observations of the electric currents traversing the earth's surface, which shew that these currents are much too weak sensibly to affect our magnetic instruments. It is not always recognised that what the magnetic observations give us is the current which actually passes from the inside to the outside of the earth. These currents are sometimes spoken of simply, as "vertical currents," and the practical man who takes his theory from "hearsay," frequently believes that he can reconcile electric and magnetic observations by imagining vertical electric currents entirely outside the earth, ascending in one region, descending in another and completed horizontally near the earth's surface, without actually cutting it. Such explanations deserve no consideration. For the present we must leave the question of vertical currents open until the general magnetic survey of the earth, undertaken by the Carnegie Institution of Washington, has furnished us with more accurate determinations in the oceanic regions.

Gauss, to whom after Gilbert we owe the groundwork of the theory of Terrestrial Magnetism, has taught us how observations, made on the surface of the earth may be used to decide whether and in what proportion the observed magnetic forces are due to causes lying inside or outside the surface. Neglecting the small effect possibly due to electric currents traversing the surface, it is always possible to imagine two independent magnetic systems which can both equally well explain the horizontal forces. Of these, one lies inside and the other outside the surface. This means that if we only observe magnetic needles swung, like a compass, so as to confine their movement to a horizontal plane, we can explain the observed deflexions equally well by forces coming to us from outside and by forces which have their seat inside the earth. But the vertical components of the forces would be quite distinct in the two cases, being in general in opposite directions. If for instance the earth were placed in a magnetic field produced by external magnets in such a way that the magnetic lines of force were all parallel as in Fig. 12, the horizontal forces on the surface of the earth would exactly be the same as those caused by a small magnet of proper intensity placed inside the earth, as illustrated in Fig. 10. In order to make the north pole at the equator point northwards, it would,

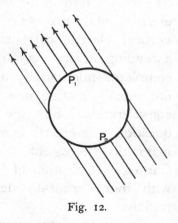

Fig. 12.

in the case now considered, be necessary to make the

lines of force run in the direction indicated by the arrows in Fig. 12, which means that at P_1 the north pole would point upwards and not downwards as is actually observed. Hence, while the horizontal forces are the same in the two cases considered, the vertical forces are in opposite directions, and this general reasoning is quite sufficient to prove that the main cause of the earth's magnetism lies inside. If however we want to decide whether any portion of it comes to us from outside, we must replace the general reasoning by an exact and troublesome calculation. It is then found that our present observations, assuming them to be correct, indicate an effect amounting to about 5 % of the total having its origin outside the earth's surface ; but here again one is bound to speak with caution until the Carnegie Institution has furnished us with accurate figures. Before leaving this portion of the subject, I must however insist on a result of Gauss' theory which is not always sufficiently recognised. There can only be one solution of the problem, and if we can explain any magnetic effect on the earth's surface by outside forces, it follows that it can not at the same time be explained by internal forces. This remark disposes of a good deal of the criticism lavished on pioneer attempts to open out this region of science. Though this criticism is often confined to a judicious shrugging of the shoulders, it stops scientific progress more effectually than active opposition, and is apt to become a constitutional habit with those who give way to its undoubted temptations.

I have hitherto spoken of magnetic forces as invariable at every point, except for the slow secular variations. In reality when a magnetic needle is delicately suspended it is never at rest, shewing that the earth's forces are always changing both in magnitude and direction. These changes are partly regular and partly irregular. Among the regular variations, that depending on the time of day is the most important. We call it the diurnal variation. It is not my intention to describe this feature of terrestrial magnetism in detail, but I should like to explain what we have learned of its origin. The greater portion of it is undoubtedly due to electric currents situated in the upper portions of the atmosphere. That the cause lies mainly above the surface can be proved, and that it lies in the atmosphere must be inferred as probable. The discussion of the problem which I gave some years ago, shewed that there is a substantial remnant which comes to us from the inside of the earth, and this is, as it should be, because the earth being a conductor, any change of the magnetic forces must give rise to induced internal currents. The intensity and phase of these internal currents allow us to draw interesting conclusions on the conductivity of the earth, but the observational data used were probably not sufficient to do more than indicate that the deeper layers must be more highly conducting than the outer shell, which is in agreement with the conclusion derived from other considerations.

If we admit that the primary cause of the diurnal variation is electro-magnetic, we may easily formulate that system of electric currents in the atmosphere, which gives the required magnetic forces. In this system the currents are everywhere parallel to the earth's surface and the question now arises whether we can explain them plausibly. The first attempt in this direction was made by Balfour Stewart, who forty years ago suggested that any motion of the atmosphere in which the particles of air are driven across the earth's lines of magnetic force would set up an electro-motive force. One great difficulty however stood in the way of this hypothesis : so far as was known at the time, air was a non-conductor requiring a considerable electro-motive force, to start a discharge and this discharge, when it did take place, was disruptive and discontinuous like a flash of lightning in a thunderstorm. It was only after I had shewn, that gases can by various means, now called processes of ionization, be put into a sensitive state in which they behave like ordinary conductors, that the main obstacle to an acceptance of Balfour Stewart's theory was removed. I have recently[1] worked out this theory so as to obtain numerical results which may be tested by observation.

If the magnetic effect be due to electro-magnetically induced currents, it should be connected directly with some diurnal change in the flow of air, and if we investigate, what must be the flow of air, which reproduce the magnetic effect, we find that it is exactly of the kind required to cause the observed diurnal change of the barometer. We are therefore led to inquire whether one and the same oscillation of the shell of air which surrounds us is sufficient to explain, on the one hand the variations of atmospheric pressure,

---

[1] *Phil. Trans.* CCVIII., p. 163 (1908).

and on the other the oscillations of the magnetic needle.

The pressure change is of two kinds, one of them reaching its highest value once a day and having therefore a period of 24 hours, while the second and more important one comes to a maximum twice in the day—like the ocean tides—having a period of 12 hours. It may at first sight seem curious, that, while the varying altitude of the sun, the main outside influence which can cause a variation in the height of the barometer, is diurnal, the oscillation of the atmosphere gives such a large preponderance to the 12 hourly change. It must be remembered, however, that the amplitude of an oscillation does not entirely depend on the intensity of the periodic force causing it, but also on the degree of agreement between the natural period of the oscillating system and that of the acting force. We may compare the oscillation of the atmosphere with that of a resonator excited by an outside sound. The resonator will, as a rule, have several independent periods in which it can freely vibrate, and ordinarily when forcibly excited its note will contain a number of these free periods, though one of them, the fundamental note, will predominate. But when forced to resonate by an outside sound coinciding in period with one of its free periods, it will respond to that period alone, the fundamental not appearing at all. It was therefore suggested first, I believe, by Lord Kelvin and then worked out more in detail by Lord Rayleigh, that the natural period of the shell of air which surrounds the earth may be nearly equal to 12 hours, and in that case the daily change of temperature, which contains a semi-diurnal variation, though its predominant period is diurnal, might set up an oscillation of 12 hours, exceeding in amplitude that of the purely diurnal oscillation. Calculations of the natural period of the free oscillations of the atmosphere, in which the effect of the rotation of the earth is taken account of, have justified this explanation.

If we calculate the motion of air, which is required to account for the amplitudes of the 24 hourly (diurnal) and 12 hourly (semi-diurnal) variation of the barometer, we find that, by properly adjusting the electric conductivity each of them may be made to explain satisfactorily the corresponding variation of the magnetic needle, but if we fix on any definite conductivity, we find too great an amplitude for the diurnal or too small an amplitude for the semi-diurnal magnetic change. We also find that the magnetic variation is much more affected by the season of the year than the barometric change. A better agreement may be obtained by making the plausible hypothesis, that the electric conductivity of the higher regions of the atmosphere is due to solar radiation, and is therefore greater in summer than in winter and also greater in day time than at night. As regards the absolute value of the conductivity, it is found that it must be about ten times as great as that which is found in the flame of a Bunsen burner. This is greater than perhaps we have any right to expect, though we know that for equal ionizing powers, the conductivity should increase with altitude. Theory would lead us to expect the conductivity to be inversely proportional to the pressure, but observations made during balloon ascents by Gerdien have led to an increase of conductivity at an altitude of six kilometres to more than ten times its normal value near the surface, while the pressure was only reduced to one half. Without wishing to lay too great a stress on the results which I have brought to your notice, they are sufficient to shew that a careful study of terrestrial magnetism may allow us to draw important conclusions on the state of the upper layers of our atmosphere.

While the normal magnetic changes which go on from day to day have received comparatively little attention, more interest has been shewn in the larger, more sudden and more irregular changes which are known as magnetic storms. This is no doubt due to the remarkable connexion between the frequency of their occurrence and the number of sun-spots which are found on the solar disc. The storms occur principally when the spots are numerous, though no certain connexion between a particular spot and a particular disturbance has yet been traced. Attempts have been made to explain the connexion, but at present they are a little more than more or less plausible guesses. It is, however, certain that we must reject the possibility of any direct magnetic action between the sun and the earth. Explanations based on such a direct effect which were never probable, were finally slain by Lord Kelvin, who shewed that the active portions of the sun's surface, if directly responsible for a magnetic storm of moderate intensity, would emit nearly four hundred times as much energy as they do in the regular supply of heat and light.

The popular view, at the present moment, is to ascribe magnetic storms to the emission by the sun of electrified particles which, passing with enormous velocity by or through our atmosphere, affect our magnetic needles and produce the storm. Neither qualitatively nor quantitatively does it seem to me that this explanation is probable, though the more refined form in which it has been presented by Professor Birkeland overcomes some of the difficulties which previously seemed fatal. Lord Kelvin's objection which, *mutatis mutandis*, seems to me to apply to this case also, must, however, for the present block the way to any theory which makes the sun responsible for the energy of magnetic storms. In a discussion of the subject[1], presented to the Royal Astronomical Society of London, I arrived at the conclusion that the store of energy residing in the earth's rotating mass is the only one sufficiently great to provide for the energy of the magnetic storm. It is large enough to do so without being sensibly affected, for after a million years, the total loss through magnetic storms would diminish the rotational velocity so little that, as a time-keeper, the earth would only lose one second per year.

I should like to explain briefly the view regarding the solar influence on terrestrial magnetism, which I have advocated for some time. I believe it to be purely an indirect one, affecting the conductivity of the upper regions of our atmosphere, either by direct radiation or by the injection of particles which ionize the air through impact. The increase in conductivity,

[1] *Monthly Notices.* R.A.S. LXV., p. 186 (1905).

in conjunction with electro-motive forces which are
always present, are, if I am right, sufficient to account
for the observed effects.

What are these electro motive forces ? We have
already discussed those due to periodic displacements
in the atmosphere, which are responsible for the
diurnal variation. There is further, that due to the
general atmospheric drift which is such as to equalise
angular momentum in different parts of the atmosphere, or from east to west in the equatorial and from
west to east in the polar regions. This drift sets up
an electro-motive force tending to drive positive
electricity towards the equator in the polar regions,
and in the opposite direction in the equatorial regions.
If the direction of motion along circles of latitude be
alone considered, and the conductivity be uniform, the
electro-motive force can be equalised by a redistribution
of electro-static force, and there is no electric current,
but if, suddenly, the conductivity be increased on one
side of the earth, a magnetic storm results which would
present in character marked similarities to that of a
certain type of magnetic storms.

Finally, there is one effect of electro-magnetic induction deserving more careful study, which has its seat
at the extreme outer limits of the atmosphere. It is
difficult to say what the nature of the induced currents
should be, because we do not know how our atmosphere ends. If it comes to an abrupt conclusion, and
if the surrounding parts of space are conducting, as
no doubt they are, induced electric currents would
pass under the influence of the rotating and magnetic
earth from the atmosphere into space, and back again.

If, on the other hand, there is a region on the out-skirts of the atmosphere which circulates with planetary velocities, the electro-magnetic induction would, roughly speaking, be in the opposite direction to that holding in the previous case. If this electro-magnetic induction be the cause of the aurora borealis, a more detailed study may decide between the alternatives, as to rest on orbital circulation of matter in space.

As the magnitude and irregularity of all our magnetic effects are exaggerated at times of maximum sun-spots, we conclude that at these times the upper atmosphere and probably also interplanetary space has an increased electric conductivity. Whether this increased conductivity of space is the cause of the solar disturbances which shew themselves as spots or their consequence, is a matter on which we are free to speculate. At any rate, if there is a variable con-ductivity of interplanetary space, it is to the magnetic needle that we must look for information, and the science of terrestrial magnetism may be expected to throw much light on whatever corpuscular matter may be contained in interplanetary space as well as on its possible relationship to the solar drift through space.

It has often been suggested that finely divided matter is being projected outwards from the sun, and the appearance of the solar corona which is a luminous envelope of the sun, visible only when the solar disc is covered by the moon in a total solar eclipse, lends some countenance to that view. We may imagine these projections to be due to electrical repulsion, or to the peculiar repulsive force which accompanies the incidence of any radiation on ab-

sorbing or scattering particles. This repulsive force
was first brought to our notice by Clerk Maxwell,
who deduced it as a theoretical consequence of his
electro-magnetic theory of light. Owing to the smallness of the effect, it seemed hopeless, at the time, to
verify it experimentally, and the theoretical calculation
was not convincing to so high an authority as Lord
Kelvin. But, as in so many cases, Maxwell proved
to be right, and we have now the experiments of
Nicols and Hull, those of Lebedew and those of
Poynting, all shewing that the propagation of radiant
energy may be considered to carry a momentum
along with it. Whenever this energy is deviated
from its straight path, or absorbed or scattered in any
way, the law of conservation of momentum gives us
at once the magnitude and direction of the resultant
repulsion, which acts whether the body on which the
radiation falls be large or small, and must even affect
the individual molecules of a gas. If the effects are
observed chiefly in the case of solid and liquid
particles, it is because the same quantity of matter
absorbs more in the solid and liquid states than in the
gaseous form. If we had a mixture of gases, a radiation transmitted through it would have a tendency to
separate the gases according to their absorbing power,
but it is questionable whether the effect would ever
become appreciable. Absorption of the more refrang-
ible rays would in general count less than absorption
in the infra-red, because most of our sources of light
and heat emit their greatest energy in the region of
low frequency. The repulsion by radiation fills an
important function in interplanetary space, explaining

possibly the appearance of comet tails and of the solar corona. It has further to be taken into consideration when discussing the stratified arrangement of substances in the solar atmosphere, but in the absence of numerical data the danger of being led astray is great.

Those who believe that the ejection of negative corpuscles explains the connexion between solar and planetary phenomena, require an occasional reminder that a body cannot indefinitely emit negative electricity without becoming charged positively, and this excess of positive electricity must be disposed of. If the sun emits negative electrons, the electric forces in its neighbourhood must be such, that there is a repulsion of positive electricity outwards. We should therefore expect a perhaps slower but equally effective discharge of positive ions, to accompany the emission of electrons.

Nothing has been said as yet on the explanation of the secular variation of terrestrial magnetism, and in our ignorance of the causes, which make the earth behave like a magnet, it is perhaps wisest to put the question aside for the present. But sometimes a hint given by the accessories of a fundamental phenomenon, throws light on the phenomenon itself, and it is therefore permissible to consider the alternatives which our present knowledge places before us. A body may act as a magnet, either because it contains magnetised iron or because electric currents circulate in it. The possibility that terrestrial magnetism may be explained by masses of magnetic iron is sometimes put aside as impossible on account of the high temperature of its interior. For iron loses its power of being permanently magnetised at about 800° C., and this temperature will be reached at comparatively moderate depths below the surface. This argument neglects the possible effects of pressure, which may considerably raise the critical temperature at which iron loses its magnetism. I have had made in the Physical Laboratory of the University of Manchester a number of experiments, in order to test whether an increase of 600 or 700 atmospheres had an effect on the magnetic properties of iron near its critical temperature. The experiments, which present very great difficulties, are not yet concluded and I am unable to give any results.

If we reject the view that iron is the cause of terrestrial magnetism, we are reduced to ascribe them to electric currents, unless we wish to bring some altogether new phenomenon into action. The objection which might be raised, that in the absence of permanent electro-motive forces such electric currents would not be permanent, but die out in consequence of the resistance of the earth, is not as serious as it looks, because Professor Horace Lamb has calculated that a system of electric currents such as is necessary to give to the earth the field of a uniformly magnetised sphere would require several million years before it is reduced to half its value. If electric currents are the cause of the earth's magnetic state, the currents producing it must have been started by some unknown cause, and their effect must die out gradually. It is perhaps interesting to note that a system of electric currents circulating in a rotating body about an axis inclined to the axis of rotation, as the earth's magnetic axis is inclined to the geographical axis would be subject to a secular variation such as is observed; but unfortunately this secular variation would be far too slow to account for the actual phenomenon.

Some years ago I made an attempt to explain the secular variation on the view that the earth contains magnetic masses of iron. If we admit for the sake of argument an inside core of the earth and an outside shell, which though connected together allow a slow relative displacement; if we further take interplanetary space to be electrically conducting and the inner core of the earth to be magnetised along an axis inclined to the axis of rotation, then magnetic forces due to electro-magnetically induced currents in interplanetary space would act on the system in such a way, that its magnetic axis would tend to become more parallel to the axis of rotation, and at the same time turn round it in the manner actually found in the secular variation.

If, ultimately, we should have to reject both the theory, that the earth's magnetism is due to masses of iron and that it is due to electric currents, we should be compelled to look for some other cause. In 1892 I suggested, and the suggestion has I think also been made by others, that every rotating body may behave as a magnet, but we have at present no experimental evidence for such a hypothesis, though it may be defended on theoretical grounds. As a suggestive argument in favour of the view, it may be mentioned, that both the main facts of terrestrial magnetism and its secular variation may be derived from the same hypothesis on the molecular constitution.

I have already mentioned in speaking of the diurnal

variation that any change in the magnetic field must be accompanied by induced electric currents in the crust of the earth. These " Earth Currents " reveal themselves whenever two metal plates are inserted into the ground and connected through a galvanometer, and at one time it was hoped that through them we might be able to learn what part they play in the observed magnetic disturbances. This hope has not been fulfilled, partly because the distance between the plates must be considerable, if errors are not to be introduced by the action of the system as a primary battery, and partly because the connexion between the earth current and the current passing through the galvanometer is not at all obvious, and depends on the electric conductivity of the soil. It must also be borne in mind, that while the magnetic effect depends on currents probably reaching down to a considerable depth, the observed earth currents are surface currents depending both in magnitude and direction on the distribution of electric conductivity in the uppermost layers of the soil.

Without entering into any details on the methods adopted for its experimental investigation, I will now describe and discuss the effects which are due to the electro-static charge which resides on the surface of the earth. The earth as a whole is charged negatively, and on the average we find that there must be nearly a million electrons on every centimetre of its surface. This negative charge is not constant, but shews irregular as well as regular diurnal and seasonal variations ; it may even be reversed in sign and reach a high value, when the sky is clouded over. The electrification of the earth, as a whole, is difficult to explain, especially when it is remembered that the atmosphere conducts electricity, and that therefore if the charge is permanently maintained it must be continually reproduced, otherwise it would quickly be dissipated.

The first question that arrests our attention is whether the earth as a whole permanently discharges negative electricity into space. This would be the case if the electro-motive force due to the charge on the surface reached the outer regions of the atmosphere. Observations made in balloons shew that the electro-static effect due to the charged earth is continually diminishing, and at heights of between 4 and 6 kilometres, it is reduced to one-tenth of its normal value at the surface. This means that the effect of the negative charge of the earth at great altitudes is counteracted by a nearly equal positive volume charge of the atmosphere. If it is reasonable to suppose that the compensation is complete in the uppermost layers, we are justified in believing that the primary cause of the earth's electrification is atmospheric and not cosmical, and that there is no dissipation into space. Though personally I share this view, I am bound to point out that the above reasoning is by no means as conclusive as it appears. Imagine a charged sphere surrounded by a medium, the conductivity of which is small near the surface of the sphere, but increases outwards. As in the steady state the quantities of electricity passing through the successive layers of the spherical envelope must be equal, it follows that the electro-motive force must be continuously diminishing outwards and a low value of that force is therefore quite consistent, with a permanent dissipation into space of all the electricity which passes from the earth into the atmosphere. In order to prove that there is no such dissipation, it would be necessary to shew that the electro-motive force diminishes with altitude more rapidly than the conductivity increases, and our observations though pointing in that direction are not yet sufficiently precise to remove the possibility of doubt.

The conductivity of the air depends mainly on its ionization, and a handy apparatus has been constructed by Ebert to count at any time or place the number of ions present. This number amounts on the average to about 1000 per cubic centimetre, there being generally an excess of positive ions present. It follows indeed from what has been said above, that an excess of negative ions would indicate either that the charge of the earth at the point of observation is reversed, or that the atmospheric conditions are such that the conductivity diminishes upwards instead of increasing as it usually does. As the negative and positive ions are continually recombining, a permanent ionizing agent must be at work. Near the surface of the earth an effective ionizer is supplied by the radio-active products escaping from the earth. At high altitudes there may be additional cosmic causes such as the ultra-violet light from the sun, but it is difficult to believe that these can penetrate through 76 centimetres of mercury, which represents the mass of the atmospheric layer above us.

The rate at which ions recombine varies considerably. Dust or mist accelerate combination, so that the number of ions, and hence the electric conductivity, depends on a balance between the rates at which they are generated and disappear. It is in agreement with this argument that mist increases the observed electric force, because it helps recombination and therefore diminishes the conductivity. Similarly Elster and Geitel, to whom together with other important contributions to the subject, we owe the first decisive proof that atmospheric air is conducting, have shewn that when the air is very transparent the electrification is considerably diminished, doubtlessly owing to the absence of dust with its recombining power.

It may, at first sight, appear strange that the electrification is not more uniform over the earth, and may even differ considerably at places which are near to each other. If we were to neglect the atmospheric charges, we should expect indeed the conductivity of the earth to be sufficient to establish equilibrium within a few seconds ; that this is not the case is itself a proof that the positive electrification neutralising the earth's negative charge resides at no great distance from the surface. This positive charge being bound to matter which can only move slowly, keeps the lines of force in position and thus prevents a uniform distribution from establishing itself quickly.

Regular observations from which the electrification of the earth may be deduced are made in many observatories, and the number of ions in the atmosphere are also being regularly counted in a few places. The diurnal and annual changes of these elements shew many features of interest, but the methods used are not well adapted to present the phenomena in their clearest light ; for, as has already been pointed out, the ionization of the air is not itself a primary phenomenon. An increased ionization may either mean that the ions are generated at a quicker rate, or that they recombine more slowly, and we shall be unable to interpret the observations which have accumulated in various observatories until these two factors are separated and we trace independently the diurnal changes in both of them.

The fundamental and, at the same time, the most difficult of all the problems of atmospheric electricity is to discover the origin of the negative charge which covers the earth. This electrification is constantly being dissipated into the air, at a rate sufficient to reduce the total charge to half its value in the course of four or five minutes, if some counteracting agent were not at work. What is the agent which continuously drives negative electricity into the earth from outside ? We might think of some surface action, or, going further afield, of some cosmical cause ; finally, we might take an intermediate view and look to the atmospheric layer of moderate altitudes as the primary separator. All three views find their advocates. The cosmic theory is rendered very improbable by the balloon observations, and such contact theories, as have hitherto been proposed, are mainly based on the observed fact that in a closed vessel containing ionized air the inner surface has a tendency to take up a negative charge, owing to the adsorption of negative ions. Fatal objections have been urged against this being a sufficient cause to explain the negative charge on the earth.

Other explanations having failed, the most promising theory seemed to be that which made the rainfall responsible for the charge. A theory put forward by Dr C. T. R. Wilson, met therefore with a very favourable reception. This theory is based on Dr Wilson's discovery, mentioned in a previous lecture, that in supersaturated air water condenses more readily on negative than on positive ions. When an upward current of moist air cools by expanding under the reduced pressure, the first condensation takes place on the particles of dust which act as nuclei, but it is likely that the upward current passing through the cloud which is thus formed, is not absolutely dry. Further condensation, the dust-free air above the cloud does not then begin until the air which is still supposed to rise becomes supersaturated; the drops formed under these circumstances are negatively electrified. If these drops collect and fall to the ground they would bring their charge to earth, the neutralising positive charge remaining in the air. Gravity here acts as separator. An experimental confirmation of this view would be obtained, if rain could be shewn to bring down an excess of negative electricity.

Let us first form an estimate of the quantities involved. Though the rainfall is only known accurately over limited regions, we may form some rough estimate of its total amount. The average all over the earth is not likely to be less than 100 or more than 200 cms. per year. If we put it at 150 cms. we shall, so far as I can judge from the published maps of rain distribution, somewhat underestimate the total amount. We may note here, in passing, how the same fact may be represented in two ways which appeal very differently to the imagination. A rainfall of 150 cms. per year means an average deposit of the twenty-thousandth part of a millimetre per second, which seems small. But summing up all over the surface of the earth it means that 26 million tons of water fall down each second of time—which creates quite a different impression. If we combine together the observed loss of charge per second of time with the quantity of water brought down, we find that for equilibrium each cubic centimetre of rain would on the average have to be charged with one-tenth of an electro-static unit of electricity.

Observations on the electric charge of rain have been first made by Messrs Elster and Geitel, and whenever similar experiments have been repeated it has been found that both positively and negatively charged rain occurs, the average charge being about one electro-static unit per cubic centimetre, which, if the charge were always of the same sign, would be ten times more than the quantity required to replenish the earth. It follows that such theories as Wilson's can only be tested by continued series of observations, determining the average excess of one charge over another. The conclusion arrived at by Elster and Geitel and some other observers, was that there is indeed an excess of negative charge.

In opposition to this are the more recent experiments made by Dr George Simpson[1], who found that

---

[1] The passage regarding Dr Simpson's experiments (*Phil. Trans.* ccix., p. 319) has been added in writing out these lectures (June, 1910). The experiments were made after the lectures were delivered.

during the monsoon the rain which falls at Simla is more often electrified positively than negatively. During the period investigated the total quantity of rain which fell was 76 cubic centimetres for each square centimetre of surface, bringing down 22·3 electro-static units of positive electricity and 7·6 of negative electricity ; there was therefore an excess of ·2 units of positive electricity for each cubic centimetre that fell. These experiments seem to have been confirmed in other localities, and should they be found to supersede the earlier experiments of Elster and Geitel, we should have to abandon what at one time appeared to be the probable source of the earth's electric charge. But, as at present the rainfall has been investigated at only a few places on the earth's surface, we must not too hastily reject the possibility that the total balance of electrification is negative although it must be conceded that the evidence at present inclines towards the opposite view.

The deadlock which for the present seems to bar our ability to explain the fundamental phenomena of atmospheric electricity, renders it advisable to look in all directions for any cause which might tend to electrify the air positively as compared to the earth.

Professor Ebert has suggested that the radio-active and therefore ionized gases, which are known to escape from the soil, precipitate some of their negative charges while filtering through the surface layers, the complementary positive ions being discharged into the atmosphere. The effect is a real one, as experiments have shown that ionized gases driven through narrow channels deposit negative charges on the walls of the passages until a definite positive charge is imparted to the escaping gas. There are, however, difficulties in the way of accepting Ebert's theory as sufficient to account for the magnitude of the earth's charge. These have been mainly urged by Dr Simpson, whose observations shew that an increased amount of radio-active gas in the air is, in general, accompanied by a diminution of the earth's charge as derived from the potential-gradient. This diminution is obviously due to the increased conductivity of the air. It would seem, therefore, that while admitting the reality of the Ebert effect, the equilibrium point would be reached with a lower electrification than that which is observed. But there are other effects tending in the same direction. Such is the spraying of sea water which, according to Lenard, imparts a positive charge to the atmosphere in which the spraying takes place. I am not aware of any observations carried on to determine the electrical condition of the atmosphere over the ocean, when its wave crests break into drops, but it is quite possible that the earth's electrification may be partly accounted for by the Lenard effect. The objection which may be raised against this and Ebert's theory, that it only accounts for the positive electrification of the air in the immediate neighbourhood of the ground, is only a superficial one, because the ordinary processes of diffusion are sufficient to cause the electrification to spread in time through the whole atmosphere.

It is not possible at present to estimate with any degree of accuracy the part played by lightning discharges in the electrical economy of the earth. F. Pockels[1] has determined in two cases the approximate maximum current in a flash of lightning,

---

[1] *Met. Zeitschrift*, 1901, p. 40.

and found it to be 10 and 20 thousand amèpres respectively. If we take the corresponding duration to be a thousandth part of a second, we should get 10 coulombs for the quantity of electricity brought down by the discharge, and we can then calculate the average number of flashes per year which would be required to charge the earth negatively to the required extent, if all flashes brought down negative electricity. I find in this way that on the average all over the earth there should be seven lightning discharges per year for each square kilometre of surface. Even making allowance for parts of the earth where thunderstorms are of almost daily occurrence, it does not seem to me, judging by present information, that lightning discharges from cloud to earth can play an important part in increasing or diminishing the charge of the earth. It would nevertheless be interesting to collect information as to the direction of the passage of electricity in a flash. Instruments intended to measure the electric condition of the earth are as a rule fine weather instruments, and are thrown off their balance during a thunderstorm, otherwise a comparison of the indications just before and after a flash would materially assist us to decide the question. That the earth is generally electrified positively when the sky is covered by rain clouds would argue in favour of the view that electricity which leaves the ground is positive.

Theories of the origin of thunderstorm electricity are almost as numerous as those accounting for the general electrification of the earth. The majority of them must be received with considerable scepticism, but Dr Simpson has recently made an important contribution

towards a rational explanation of what goes on in a
thunderstorm. He has proved that, contrary to what
was hitherto believed, the splashing and breaking up of
drops of water shew the "Lenard" effect, that is to say,
cause the drops to be charged positively while negative
electricity is dissipated into the air. A thunder cloud
is known to form in a rapidly ascending current of air.
Lenard, in an important research, has shewn that a
drop of water as it falls, never reaches a velocity
greater than 8 metres per second however large the
drop, while drops having a diameter of 1·5 millimetres
fall with a velocity of about half that amount. An
ascending current of 8 metres per second will therefore keep the largest drops stationary in suspension,
while the smaller drops will be carried upwards. The
larger drops break up in the air, and doing so become
positively electrified according to Dr Simpson. If the
ascending current spreads out laterally near the top of
the cloud, the vertical velocity is diminished, the drops
will grow and fall, but only to break up and be carried
upwards again. A quantity of electricity large enough to
account for the lightning discharge can thus accumulate
in a cloud. Lightning can pass from the positively
charged raindrops to the negative charge, which is
carried away to the ascending and spreading current
of air, or from either of the charges to earth.

The interesting position in which the science of
atmospheric electricity is placed at the present moment,
must be my excuse if I have treated the subject with
a greater amount of detailed attention than I have
given to other parts of terrestrial physics.

Before concluding this course of lectures, I feel that

I ought to refer to a question on which we thought we knew something, until the discovery of radio-activity threw us back to the primitive state where no opinion is absurd and every hypothesis permissible. That is the question of the age of the earth as an inhabitable globe. The calculations of Lord Kelvin were based on the known laws of conduction of heat, and on the observed increase in temperature from the surface of the earth downwards. Taking the reasonable view that the earth originally was fluid and cooled down by radiation into space, a time must have been, when the crust began to solidify. According to Lord Kelvin, solidification then spread through the mass and no further diminution of temperature took place until the whole earth was solidified. From known data, among which the one concerning the temperature of solidification is the most uncertain, he estimated that about 50 million years must have lapsed since the solidification was complete. Geologists however were not satisfied with the shortness of time allowed to them by Lord Kelvin for getting their strata in order and demanded at least double the time. They may have it now, and as much more as they desire, because the generation of heat supplied by the decay of radio-active products may alone be sufficient to account for the loss of heat by radiation. So far as the laws of heat can supply us with information, the earth may have been in its present state during an infinite time, and Lord Kelvin's estimate now stands as the smallest allowable time, consistent with these laws. We owe the recent investigation of this subject to the

Honourable R. J. Strutt[1], who measured the radium
contents of a large number of minerals, and found that
if the interior of the globe contained as much radium in
proportion to its volume, as its surface layers, the heat
generated would far exceed the loss of radiation from
the surface. The earth would therefore get hotter
and not colder with time. If there is equilibrium of
heat, the loss through radiation being counterbalanced
by the generation of heat in the outer crust which
surrounds an inner core assumed to be radium free,
Strutt finds for the outer crust a thickness of 45 miles,
and for the inner nucleus a temperature of about
1500° C. These calculations must alter our views
materially, though they are based on the assumption
that the heat production of radium is not affected by
a temperature of 1500° C., and a pressure which is considerably higher than anything we can produce in our
laboratory. This is a reasonable assumption, because
so far as we are able to judge at present, the rate of
decay of radio-active products is entirely independent
of external circumstances.

Lord Kelvin's conclusions regarding the solidity
and rigidity of the earth stand however firmer to-day
than ever. He based his opinion on the phenomena
of precession and nutation and found the observed
facts to be inconsistent with the idea of a liquid
nucleus. This result has quite recently received a
striking confirmation by the measurements of Hecker
of the actual deformation of the earth consequent on

[1] *Proc. Royal Soc.* LXXVII. p. 472 (1906).

tidal action[1]. These observations give us definite information that the rigidity of the earth must be about the same as that of steel. This is in complete agreement with results obtained by other methods, the most direct of which is perhaps that deduced from the rate of propagation of earthquake waves. When dislocation of matter in the earth's crust takes place with sufficient suddenness to cause a tremor all over the surface of the earth, the disturbance is propagated by means of waves, and we distinguish three types of waves. One is the wave of compression, identical in character with the sound wave ; the second is a wave of distortion resembling more nearly an electro-magnetic wave, while the third is a wave running along the surface like the waves of the sea. The wave of distortion cannot exist in a liquid body, and as we find that it actually does pass right through the central portions of the earth, we conclude that the earth must be solid throughout, except perhaps in isolated regions. The rate of propagation gives, if the density of material be known, a value for the rigidity, and the number that is found in this way is quite consistent with Hecker's result. But the most curious of all proofs of the great rigidity of the earth is furnished by the movement of the earth's poles. The earth's axis, according to astronomical observations, changes its position with reference to the earth itself, so that the north pole is not a fixed point. The displacement is very small, the position of its poles always falling within a circle of about 10 metres radius. It has been known for a long

---

[1] This only became known to me after these lectures were delivered.

time that theoretically such a "wabble" was possible ;
but it remained undiscovered, because, the theoretical
calculation which assumed the earth to be a perfectly
rigid body, gave 10 months for the period of revolution of the axis, and it was this period that was looked
for. It was only comparatively recently that Chandler
discovered a period not of 10 months, but of 14
months, and Newcomb proved that the difference
between 10 and 14 months could be explained if the
earth were yielding, but only so much as a body
having approximately the rigidity of steel would do.
It is always satisfactory when we find that different
lines of reasoning lead to the same result, and at
present there is hardly one, in the domain of Geo-
Physics which stands on so firm a basis as that giving
to the earth an extremely high rigidity.

In concluding these lectures I realise, that I have
omitted many important subjects and none more so than
that universal force of nature which at present stands disconnected from all other forces : the force of gravitation.
Experiments were made by Michael Faraday, than
whom no one was better qualified, with the object of
finding some relationship between gravitation and other
effects of nature, but without result. Many theories no
doubt have been proposed, and one, that of Lesage,
has received some support, until Maxwell shewed it to
be untenable. It may be revived some day in a less
vulnerable form, but for the present we may dis-
regard it.

Lorentz suggested that if the repulsion between
two quantities of electricity of the same kind is a little
less than the attraction between the same quantities of

different kinds, then adopting the electron theory of matter, gravitation might be explained. Another view which has always fascinated me, is based on the remark made by Lord Kelvin, that two sources of fluid, in other words two points of space from which a liquid flows out in all directions, attract each other according to the law of gravitation. Now is it possible that all matter, whether built up of electrons or not, discharges particles indefinitely in all directions, thus acting like a source of fluid? Our first impulse, no doubt, would be to reject this view as untenable, for it would result in a gradual dissipation of all matter. But if all forces were to diminish in the same ratio as the masses, we might live in a universe in which matter was constantly destroyed, without our ever noticing it, because however little were left, the effects depending on the ratio of force to mass would remain the same. But while recognising the possibility of a continual dissipation of mass, we are not necessarily forced to admit it, if we adopt the above theory of gravitation. The universe must have begun by a process which lies outside physical laws, and it seems to me no easier to grasp the conception of a creation which took place at one single time than a creation which continues throughout all ages. Indeed, if we come to think of it, the continuance of a physical law like that of gravitation is as much a miracle as the continuous uniform creation of matter would be.

One final thought in conclusion. We recognise atoms of matter, such as atoms of helium or of hydrogen, and we can observe these atoms in stars so far away that the light from them takes several hundred

years to reach us.   Now how far are these molecules of
the same substance identical with each other?   If you
weigh measurable quantities, containing millions and
millions of them, you get the same weight to a fraction,
but that does not prove identity between individuals.
You might as well argue, that if the total weight of
two flocks of sheep, containing the same number, is
the same, therefore all individual sheep must be equally
heavy.   Molecules might differ from each other when
weighed singly, but when weighed by the million the
total might agree, if there is a certain average weight,
which they all more or less approach.   Now spectrum
analysis allows us to discover discrepancies between
individual molecules, and though we cannot speak
absolutely, because our instruments are not perfect,
we find that the light which is emitted by one sample
of helium, which may be on a star, is so nearly equal
to that of another sample on the earth, that there is
no room for the possibility of any great differences
between individual molecules.   We may form a numerical estimate of possible discrepancies between the
molecules of two samples of terrestrial helium, by
assuming the observed want of homogeneity of spectrum
lines to be caused by real differences in the periods of
vibration.   This of course gives us an outside value.
To put the result in a form, which is readily appreciated, I will compare different molecules to a
number of clocks, the period of vibration of the
molecule corresponding to the time of swing of the
pendulums.   If the clocks were rated to the same
accuracy as the molecules are found to be, there would
after 23 days running be no more than 10 of them,

shewing an error of as much as one second. It is more than probable that increased powers of observation will prove an even nearer approach to identity in the periods of molecular vibration! This close agreement between the masses of molecules of the same kind, and between the forces which regulate the luminous motion of the electrons is an essential condition to be satisfied by any theory of atomic structure.

INDEX

Index 163

Printed in the United States
By Bookmasters